FM 21-15

US WAR DEPARTMENT BASIC FIELD MANUAL

AF730875

EQUIPMENT, CLOTHING, AND TENT PITCHING

1940

WORLD WAR II

CIVILIAN REFERENCE EDITION

UNABRIDGED HISTORIC ARMY HANDBOOK ON STANDARD ISSUE EQUIPMENT FOR SOLDIERS AND PROPER USE OF PERSONAL SUPPLIES

U.S. WAR DEPARTMENT

Doublebit Press

New content, introduction, cover design, and annotations:
Copyright © 2021 by Doublebit Press. All rights reserved.
www.doublebitpress.com | Cherry, IL, USA

Original content under the public domain; unrestricted for civilian distribution.
Originally published in 1940 by the U.S. War Department.

A Part of the Military Outdoors Skills Series

Doublebit Press Civilian Reference Edition ISBNs
Hardcover: 978-1643892160
Paperback: 978-1643892177

WARNING: Some of the material in this book may be outdated by modern safety standards. This antique text may contain outdated, unsafe, or now illegal activities, projects, or mechanical, electrical, chemical, or medical practices and operations. Any use of this book for purposes other than historic study may result in unsafe and hazardous conditions and individuals act at their own risk and are responsible for their own safety. Doublebit Press, its authors, or its agents assume no liability for any injury, harm, or damages to persons or property arising either directly or indirectly from any content contained in this text or the activities performed by readers. Remember to be safe with any activity or work you do, use good judgement by following proper health and safety protocols, and be properly trained. In addition, because this book was from a past time and is presented in an unabridged form, the contents may be culturally or racially insensitive. Such content does not represent the opinions or positions of the publisher and are presented for historical posterity and accuracy to the original text.

DISCLAIMER: Doublebit Press has not tested or analyzed the methods, materials, and practices appearing in this public domain text and provides no warranty to the accuracy and reliability of the content. NOT CURRENT MILITARY DOCTRINE: no military operations or methods described within are endorsed or suggested by the publisher and this text is not intended to be used in any actual operations or programs. This text is published for entertainment, historical study, and personal literary enrichment purposes only and should only be used for such. This text is provided only as a reprinted facsimile from the unedited public domain original as first published and authored by the original author and publisher. The publisher assumes no liability for any injury, harm, or damages to persons or property arising either directly or indirectly from any information contained in this public domain book or activities performed by readers.

FM 21-15

BASIC FIELD MANUAL

EQUIPMENT, CLOTHING, AND TENT PITCHING

Prepared under direction of the
Chief of Staff

UNITED STATES
GOVERNMENT PRINTING OFFICE
WASHINGTON: 1940

For sale by the Superintendent of Documents, Washington, D. C. - Price 15 cents

WAR DEPARTMENT,
WASHINGTON, *October 1, 1940.*

FM 21–15, Equipment, Clothing, and Tent Pitching, is published for the information and guidance of all concerned.

[A. G. 062.11 (6–28–40).]

BY ORDER OF THE SECRETARY OF WAR:

G. C. MARSHALL,
Chief of Staff.

OFFICIAL:

E. S. ADAMS,
Major General,
The Adjutant General.

TABLE OF CONTENTS

BASIC FIELD MANUAL

EQUIPMENT, CLOTHING, AND TENT PITCHING

(The matter contained herein supersedes Basic Field Manual, Volume 1, Chapter 3, December 31, 1937; TR 225–5, July 10, 1925 (including C1, January 3, 1927); TR 225–10, July 10, 1925 (including C1, January 2, 1931); and TR 225–15, January 26, 1926.)

SECTION I

GENERAL

■ 1. GENERAL.—The allowances of clothing and equipment referred to in this manual are prescribed in Tables of Basic Allowances. Procedure of issue, fitting, marking, accounting, and disposition is prescribed in Army Regulations. The care of arms and technical equipment is prescribed in the appropriate Field Manuals and Technical Manuals. The articles constituting the uniform and the wearing of the uniform are prescribed in Army Regulations.

■ 2. SCOPE.—This manual prescribes the wearing, packing, and display for inspection of the field equipment pertaining to mounted and dismounted personnel; also the care of leather and web equipment used in the field.

■ 3. RESPONSIBILITY FOR EQUIPMENT.—The individual is primarily responsible that the field equipment issued to him is maintained in condition for field service, and that the proper finishes of metal, leather, and painted articles are protected or renewed. Company commanders will supervise the renewal of worn surfaces and refinishing thereof.

■ 4. CARE OF LEATHER.—*a. Cleaning and oiling agents.*—Two agents are essential to the proper care of leather; a cleaning agent and an oiling agent. Castile or other mild soap is the best cleaning agent, while neat's-foot oil meets the requirements for oiling. The composition of saddle soap is such that it both cleans and oils. Dubbin is particularly suited for use on shoes and boots in the field because it contains ingredients that waterproof as well as preserve the leather.

b. Boots and shoes.—(1) Dirt or mud will be removed from boots and shoes by brushing or by scraping with a dull instrument. They will then be cleaned with soap. Drying by direct exposure to sun, fire, or strong heat is injurious to the leather and is prohibited.

(2) Shoes will be well broken in and maintained in serviceable condition.

(3) In garrison, regimental, separate, or higher unit commanders may prescribe that boots and shoes be polished.

c. Harness and saddlery.—(1) *Daily care.*—Leather equipment in use will be cleaned daily. Under no conditions will it be cleaned by immersion in water.

(2) *Procedure in cleaning.*—(*a*) Separate all parts, unbuckle straps, and remove detachable parts permissible.

(*b*) With a damp (not wet) sponge remove all surface dust, mud or other foreign substances.

(*c*) Rinse the sponge and, using very little water, make a heavy lather by rubbing it vigorously on the soap or saddle soap. When a creamy lather is obtained, clean each piece of equipment thoroughly. Each strap is drawn its entire length through the lathered sponge to remove the dirt and sweat from the leather.

(*d*) After the leather has become partially dry, rub it vigorously with a soft cloth.

(*e*) Generally, the above will be sufficient to keep the leather in excellent condition. However, at intervals it may be necessary to oil very lightly with neat's-foot oil. In doing this, care should be taken to apply the oil sparingly on the flesh or unfinished side of the leather, using light quick strokes in order to obtain even distribution.

(*f*) Excessive oil is injurious to leather equipment. Oil is applied sparingly at time intervals that can only be determined by an examination of the appearance and pliability of the leather.

■ 5. CARE OF WEB EQUIPMENT.—Web equipment will be washed with castile or similar mild soap and dried in the *shade*. Use of yellow issue soap, or of commercial cleaning fluids or dyes that may injure component metal parts and leather is prohibited.

■ 6. CARE OF METAL PARTS.—Bits, curb chains, and all metal parts issued unpainted will be oiled lightly when not in use. When in use, they will be kept clean and free from rust. Except for authorized renovation, the removal of paint from metal parts which are issued painted is prohibited.

■ 7. CLOTHING.—*a. Basis of issue.*—(1) Outer articles of uniform clothing for all grades are listed and described in AR 600–35. Authorized articles of clothing are issued to Army nurses and enlisted men in accordance with the amounts prescribed in the Table of Basic Allowances pertaining to the respective arm or service concerned. Normally, officers and warrant officers are required to provide themselves with the prescribed articles of uniform and other clothing. In the theater of operations, officers and warrant officers may be issued articles of clothing not exceeding the amounts prescribed in Tables of Basic Allowances.

(2) Organization commanders are responsible that the authorized allowance of clothing is issued to personnel of their commands, that the clothing is marked as required, and that it is used and cared for and records kept of it as prescribed by regulations. (See AR 600–35, 615–40, and 850–5.)

b. The uniform.—Organization commanders are responsible for the proper fit of the clothing of their commands. (See AR 600–35.) It is especially important that enlisted men's socks and shoes are of correct size. The method prescribed in AR 850–125 for the fitting of shoes and socks will be followed. The uniforms of officers and enlisted men will conform, both on and off duty, in every respect to the specifications therefor and will be worn in the prescribed manner. (See AR 600–35 and 600–40.) In particular, the following points will be observed in wearing the uniform:

(1) The uniform will be kept clean and neat and in good repair.

(2) Missing insignia and buttons will be promptly replaced.

(3) Insignia and buttons on the service uniform during

peace will be of a bright finish; in war, they will have a dull bronze finish.

(4) Insignia and buttons having a gold finish should be cleaned with ammonia and water. An abrasive will remove gold plating. Rubber, manila paper, or any material containing surphur if near medals, insignia, or buttons, will tarnish them.

(5) Overcoats, coats, and shirts will be worn buttoned throughout.

(6) Service hats are worn in regulation shape and the hat cord will be sewed on. (See AR 600–35 and 600–40.)

(7) Field caps of standard design are worn by personnel of the Air Corps and of mechanized and motorized units, by all drivers of motor vehicles, and by personnel assigned to motor vehicles. (See AR 600–35 and 600–40.)

■ 8. INSIGNIA OF GRADE.—Insignia of grade on the service uniform are worn by officers on each shoulder loop, by warrant officers on the shirt collar or coat lapel, and by noncommissioned officers and privates, first class, on each sleeve as illustrated in figure 1.

■ 9. HEADGEAR INSIGNIA AND HAT CORDS.—*a. Headgear insignia.*—See figure 2.

b. Hat cords.—(1) *General officers.*—A double cord of gold bullion with an acorn of same material.

(2) *Other officers.*—A double cord of gold bullion and black silk intermixed, the acorns and keeper of gold bullion.

(3) *Warrant officers.*—A double cord of silver and black silk intermixed, the acorns and keeper of black silk.

(4) *Enlisted men.*—A double cord of the color of the arm or service. When two colors are prescribed for an arm or service, the cord will be in the color first named and the acorns and keeper will be in the second named.

■ 10. BRASSARDS.—Brassards consist of distinctive bands of cloth worn around the left sleeve between the elbow and the shoulder and are for the ready identification of persons detailed to special duties. They should be about 4 inches in width and about 18 inches long.

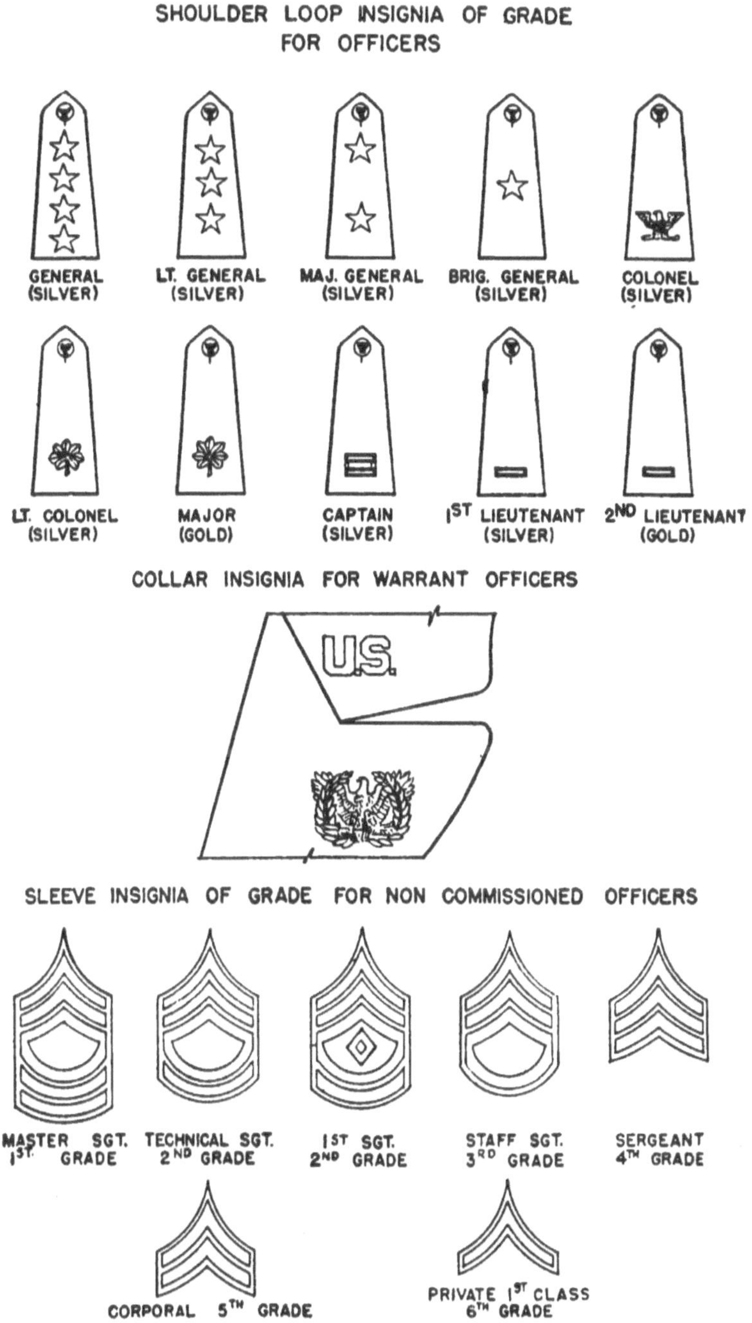

FIGURE 1.—Insignia of grade.

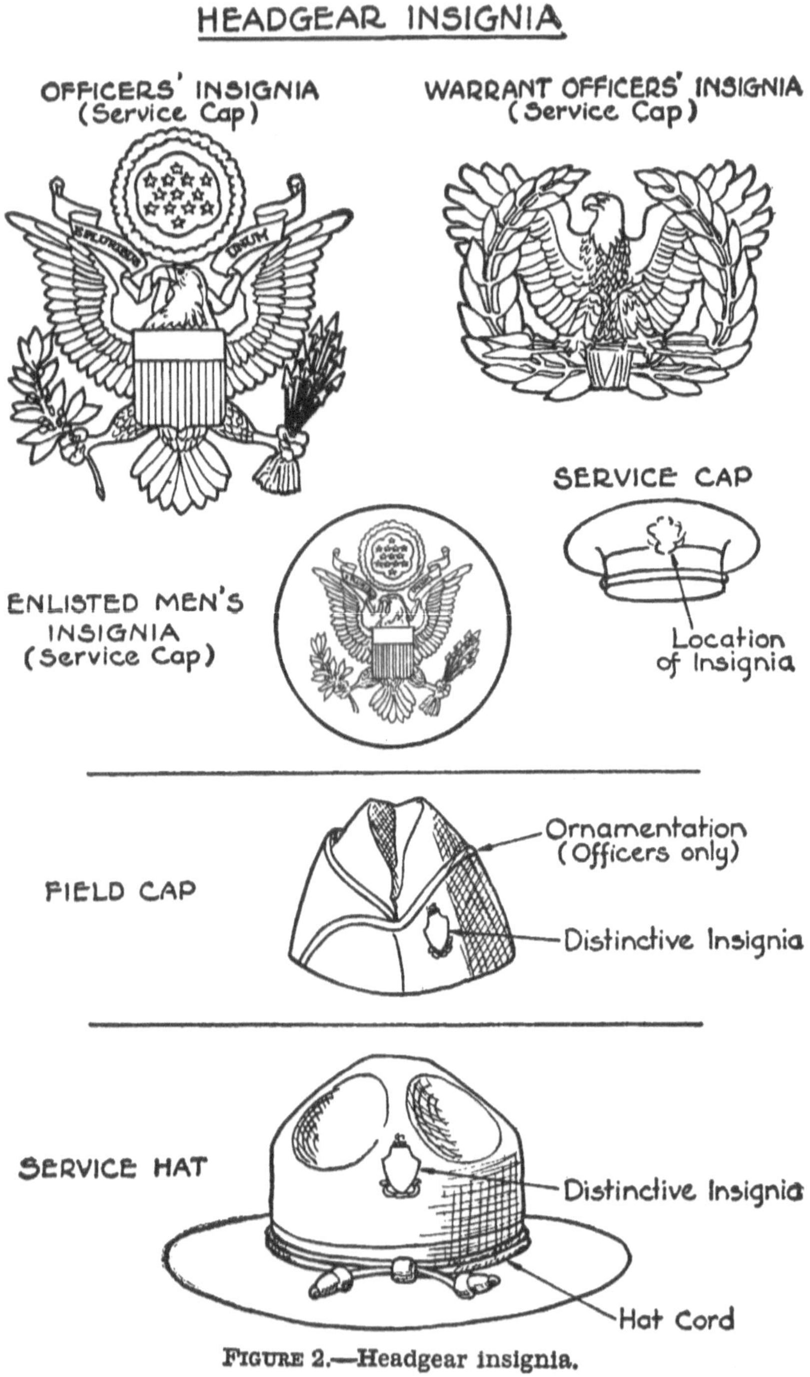

FIGURE 2.—Headgear insignia.

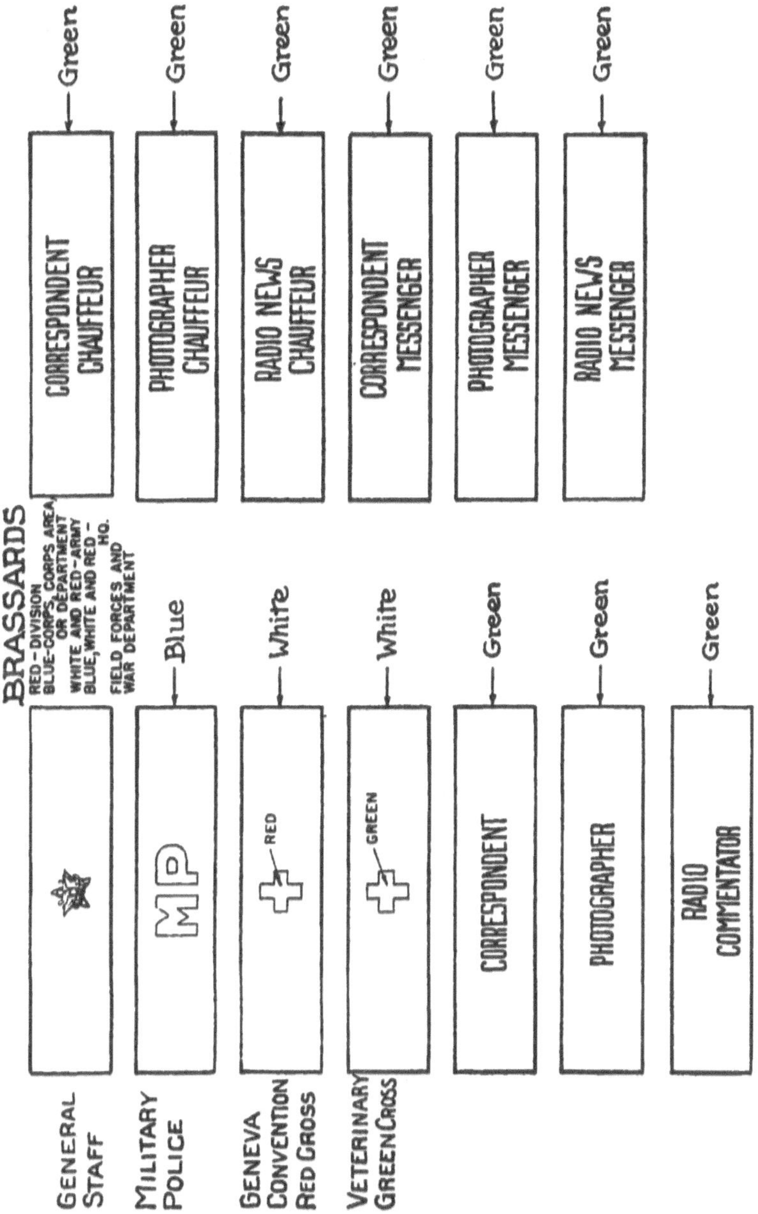

FIGURE 3.—Brassards.

Section II

OFFICERS' AND WARRANT OFFICERS' EQUIPMENT

■ 11. General.—The equipment covered in this section is the *basic* equipment common to the greater portion of the service. Other special equipment peculiar to certain arms and services or particular conditions will be carried in the most practicable manner possible.

■ 12. Method of Carrying.—For normal methods, see Table I.

TABLE I.—*Officers' and warrant officers' field equipment (other than clothing worn on person).*

Article	Dismounted	Mounted on horse [1]	Mounted on vehicle
Bag, cantle		Attached to rear of saddle.	
Bag, canvas, field	Attached to suspenders, carried on back, or on right side slung by strap passing over left shoulder.		Attached to suspenders, carried on back; on right side slung by strap passing over left shoulder or stowed in vehicle.
Bags, feed and grain		On pommel under raincoat or overcoat.	
Basin, canvas, folding	In bedding roll	In bedding roll	In bedding roll.
Belt, pistol or revolver	Worn [2]	Worn [2]	Worn.[2]
Blanket, wool	In bedding roll	In bedding roll	In bedding roll.
Boots or shoes	In bedding roll	In bedding roll	In bedding roll.
Breeches	In bedding roll	In bedding roll	In bedding roll.
Bucket, canvas folding	In bedding roll	In bedding roll	In bedding roll.
Canteen, cup and cover	On belt, left rear	Off (right) cantle ring	On belt, left rear.
Case, dispatch	On right side, slung by a strap passing over left shoulder.	On right side, slung by a strap passing over left shoulder.	On right side, slung by a strap passing over left shoulder.
Compass, prismatic	Right front of belt	Right front of belt	Right front of belt.
Flashlight	In bag, canvas, field	Near (left) cantle pocket	In bag, canvas, field.
Glasses, field	On right rear slung by strap passing over left shoulder.	On right rear slung by strap passing over left shoulder.	On right rear slung by strap passing over left shoulder.
Handkerchief	In bag, canvas, field and bedding roll.	In near (left) saddle bag	In bag, canvas, field and bedding roll.

[1] Saddle load may be distributed between first and second mount when second mount is available.
[2] See paragraph 13*b*.

TABLE I.—*Officers' and warrant officers' field equipment*—Con.

Article	Dismounted	Mounted on horse	Mounted on vehicle
Helmet, steel	On back, attached to bag, canvas, field.	Attached to near (left) saddle bag	On back, attached to bag, canvas, field.
Holster, pistol	On belt opposite right hip	On belt opposite right hip	On belt opposite right hip.
Horseshoes (2), and nails		In near (left) pommel pocket	
Kit, mess, complete	In bag, canvas, field	Near (left) cantle bag	In bag, canvas, field.
Laces, extra	In bag, canvas, field	In off (right) cantle bag	In bag, canvas, field.
Mask, gas, service	Under left arm, slung by strap passing over right shoulder.	Under left arm, slung by strap passing over right shoulder.	Under left arm, slung by strap passing over right shoulder.
Mask, gas, horse		Attached to halter under throat latch.	
Overcoat	On bag, canvas, field	On cantle over feed bag	On bag, canvas, field.
Pocket, magazine, web, double.	Left front of belt	Left front of belt	Left front of belt.
Pockets, pommel, pair		One attached to each side of pommel.	
Pouch, first-aid packet	On belt, right rear	On belt, right rear	On belt, right rear.
Raincoat	In bag, canvas, field	On cantle	In bag, canvas, field.
Rations	In bag, canvas, field	Distributed between cantle pockets to balance load.	In bag, canvas, field.
Roll, bedding	On cargo vehicle (or on pack animal).	On cargo vehicle (or on pack animal).	On cargo vehicle.
Saddle soap and sponge		In near (left) pommel pocket	
Set, toilet	In bag, canvas, field	In off (right) pommel pocket	In bag, canvas, field.
Shirt, flannel	In bedding roll	In bedding roll	In bedding roll.

Socks, pair	In bedding roll	In off cantle bag	In bedding roll.
Surcingle		In near pommel pocket	
Suspenders, belt	Worn, attached to belt [2]	Worn, attached to belt [2]	Worn, attached to belt.[2]
Tag, identification, with tape.	Around neck under shirt	Around neck under shirt	Around neck under shirt.
Tent, shelter (2 halves), complete.	In bedding roll	In bedding roll or on second mount.	In bedding roll.
Towel	In bag, canvas, field, or in bedding roll.	In off (right) pommel pocket	In bag, canvas, field, or in bedding roll.
Underclothing	In bedding roll	In off cantle bag	In bedding roll.
Whistle (chain, hooked to left shoulder loop buttonhole).	Left pocket, shirt or coat	Left pocket, shirt or coat	Left pocket, shirt or coat.

[2] See paragraph 13*b*.

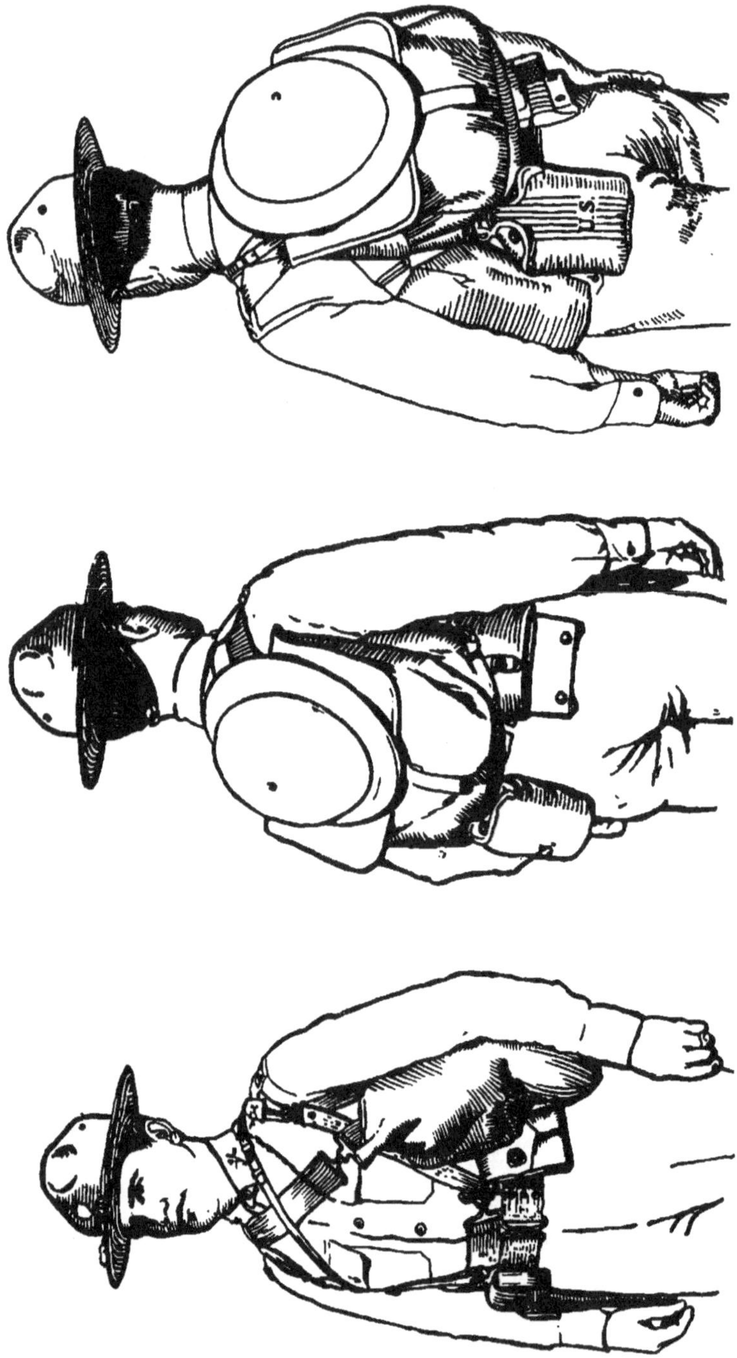

(1) Front view. (2) Right rear. (3) Left rear.

FIGURE 4.—Dismounted officers' field equipment.

■ 13. Method of Assembling.—*a. Belt.*—To assemble, place belt on the ground with outer side of belt down. Insert hook end of belt through sliding keeper, then through unattached male buckle with stud down, double belt back through sliding keeper and secure end hook in the particular center eyelet required to give proper length. Slide magazine pouch over attached female buckle end and attach it to the fastener provided. The belt is put on with male buckle on the man's right.

b. To attach suspenders to belt.—Lay belt on the ground with articles attached, inner side of belt down. Attach snap hooks of rear suspender straps to two of the upper row of eyelets, respectively, to right and left of center of belt, point of hooks out. Attach snap hooks of right and left front suspender straps to right and left ends of belt in a similar manner so that the front strap on each side will be attached to first eyelet from buckle and the other strap to third eyelet from buckle.

c. To attach field bag to suspenders.—Place assembled pistol belt and suspenders on the body and make necessary adjustment as to length of straps. With field bag suspended by the hands in front of the body, outer side of bag toward the body, attach snap hooks, point in, to suspender D-rings. Pass bag over the head to its position on the back. Position on the back of wearer may be varied by regulating front and rear suspender adjustment. Disengage the rearmost of front suspender straps on each side from belt and attach them to the D-rings at lower, inner corners of field bag. Access to field bag may be had by disengaging snap hooks from D-rings of bag and passing bag over the head to front of body.

d. Helmet.—Secure by looping chin strap about field bag suspension straps so that top of helmet is outward (See fig. 4②.)

Section III

ENLISTED MEN'S EQUIPMENT

■ 14. General.—The equipment covered in this section is the *basic* equipment common to the greater portion of the service. Other special equipment peculiar to certain arms and services or particular conditions will be carried in the most practicable manner possible.

■ 15. Method of Carrying.—For normal method, see Table II.

TABLE II.—*Field equipment, enlisted men (other than clothing worn on person)*

Article	Dismounted	Mounted on horse (except artillery drivers)	Driver, horse (artillery only)	Driver, vehicle	Men mounted in vehicle [1]
Bag, canvas, field, with carrying strap.	On right side, slung by a strap passing over left shoulder.	----	----	On right side slung by a strap passing over left shoulder; or in/on vehicle.	On right side, slung by a strap passing over left shoulder; or in/on vehicle.
Bags, feed and grain.	----	On pommel under raincoat.	On seat of saddle, off horse, or on limber.	----	
Belt, pistol, revolver, cartridge, or magazine.	Worn [2]	Worn [2]	Worn [2]	Worn [2]	Worn.[2]
Blanket, wool	In pack carrier. Carried on back, or in cargo vehicle.	In cantle or blanket roll.	In cantle or blanket roll.	In pack carrier or in blanket roll. Carried on back or in/on vehicle.	In pack carrier or in blanket roll. Carried on back or in/on vehicle.
Canteen, cup and cover.	On belt, left rear	Slung from off (right) cantle ring and attached to off saddlebag.	Slung from near (left) cantle ring, off horse and attached to saddlebag.	On belt, left rear or in/on vehicle.	On belt, left rear or in/on vehicle.
Glasses, field	On right side, slung by strap passing over left shoulder.	On right side, slung by strap passing over left shoulder.	----	----	On right side, slung by strap passing over left shoulder.
Handkerchief	In blanket roll	Near (left) saddlebag.	Near (left) saddlebag off horse.	In blanket roll, or in saddlebag.	In blanket roll.

Harversack	On back attached to belt.[2]			On back attached to belt.[2]	On back attached to belt.[2]
Helmet, steel	Attached to rear of haversack.	Attached to near (left) saddlebag.	Attached to off (right) saddlebag, off horse.	Attached to bag, canvas, field, near saddlebag, or to rear of haversack.	Attached to haversack, or bag, canvas, field.
Holster, pistol	Attached to belt opposite right hip.	Attached to belt opposite right hip.	Attached to belt opposite right hip.	Attached to belt opposite right hip.	Attached to belt opposite right hip.
Horseshoes (2), with nails.		In off (right) saddlebag.	In off (right) saddlebag, off horse.		
Intrenching tool (machete or bolo).	Attached to rear of haversack.				Attached to rear of haversack.
Kit, grooming, complete, and saddle soap and sponge.		In off (right) saddlebag.	In off (right) saddlebag, off horse.		
Kit, mess, complete	In haversack	In near (left) saddlebag.	In near (left) saddlebag, off horse.	In bag, canvas, field, in haversack, or in saddlebag.	In haversack or in bag, canvas, field.
Laces, extra	In haversack	In near (left) saddlebag.	In near (left) saddlebag, off horse.	In bag, canvas, field, in haversack, or in left saddle bag.	In haversack or in bag, canvas, field.
Mask, gas, horse		Strapped to halter under throat latch.	Strapped to halter under throat latch.		
Mask, gas, service	Slung under left arm by strap passing over right shoulder.	Slung under left arm by strap passing over right shoulder.	Slung under left arm by strap passing over right shoulder.	Slung under left arm by strap passing over right shoulder.	Slung under left arm by strap passing over right shoulder.

[1] Alternative methods for carrying equipment of men mounted in vehicles are prescribed for the reason that types of vehicles, nature and lengths of march, etc., vary so greatly that the description of only one method for one type of vehicle will not suffice.

[2] See paragraph 16.

TABLE II.—*Field equipment, enlisted men (other than clothing worn on person)*—Continued

Article	Dismounted	Mounted on horse (except artillery drivers)	Driver, horse (artillery only)	Driver, vehicle	Men mounted i vehicle[1]
Overcoat	Attached to haversack.	Strapped to pommel over feed bag.	Strapped across seat of saddle, off horse.	Attached to haversack, to bag, canvas, field, or in/on vehicle.	Attached to haversack, to bag, canvas, field, or in/on vehicle.
Pocket, magazine, web, double.	Left front of belt	Left front of belt	Left front of belt	Left front of belt	Left front of belt.
Pouch, first-aid packet.	On belt, right rear	On belt, right rear	On belt, right rear	On belt, right rear	On belt, right rear.
Raincoat	In haversack	Attached to pommel over feed bag.	Strapped across seat of saddle, off horse.	In haversack, in bag, canvas, field, or in/on vehicle.	In haversack, in bag, canvas, field, or in/on vehicle.
Rations	In haversack	Distributed between the two saddlebags to balance load.	Distributed between the two saddlebags off horse, to balance load.	In haversack, in bag, canvas, field, or in saddlebags.	In haversack, or in bag, canvas, field.
Saddlebags, pair		On bars of saddle in rear of cantle.	Across seat of saddle or on bars of saddle in rear of cantle, off horse.	Motorcyclists—i rear of seat.	
Scabboard, bayonet	Attached to left side of haversack.				Attached to left side of haversack.
Scabbard, rifle		Attached to near (left) side of saddle, under skirt.		Attached to vehicle	

Set, toilet	In haversack	In near (left) saddlebag.	In near (left) saddlebag, off horse.	In haversack, in bag, canvas, field, or in saddlebag.	In haversack, or in bag, canvas, field.
Socks, pair	In blanket roll	do	do	In blanket roll or in saddlebag.	In blanket roll.
Surcingle		In off (right) saddlebag, or over saddle blanket.	Attached, one to each saddlebag, off horse.		
Suspenders		Attached to belt	Attached to belt	Attached to belt	Attached to belt.
Tag, identification, with tape.	Around neck, under shirt.	Around neck, under shirt.	Around neck, under shirt.	Around neck, under shirt.	Around neck, under shirt.
Tent, shelter half complete with pole, rope, and pins.	In pack carrier, carried on back, or in vehicle.	Covering (pole, rope, and pins within) blanket roll, on cantle or in/on vehicle.	Covering (pole, rope, and pins within) blanket roll, carried on seat of saddle, off horse, or on limber.	Covering (pole, rope, and pins within) blanket roll, carried in/on vehicle.	Covering (pole, rope, and pins within) blanket roll. Carried on back or in/on vehicle.
Towel, face	In haversack	In near (left) saddlebag.	In near (left) saddlebag, off horse.	In haversack, in bag, canvas, field, or in saddlebag.	In haversack or in bag, canvas, field.
Trumpet	Suspended on right side by strap passing over left shoulder.	Suspended on right side by strap passing over left shoulder.			Suspended on right side by strap passing over left shoulder.
Underclothing	In blanket roll. Carried on back or in cargo vehicle.	In near (left) saddlebag, or blanket roll.	In near (left) saddlebag, off horse, or blanket roll.	In blanket roll, in bag, canvas, field, or in saddlebag.	In blanket roll. Carried on back or in cargo vehicle.
Whistle (chain, hooked to left shoulder loop buttonhole).	In left pocket, shirt or coat.	In left pocket, shirt or coat.	In left pocket, shirt or coat.	In left pocket, shirt or coat.	In left pocket, shirt or coat.

■ 16. Method of Assembling Haversack and Pack Carrier.—*a. Belt.*—(1) *Pistol or revolver belt* (fig. 5).—Place belt on the ground, outer side of belt down. Insert hook end of belt through sliding keeper, then through unattached male buckle

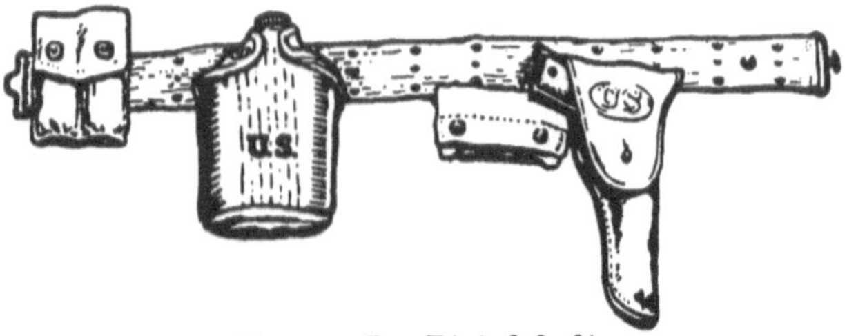

Figure 5.—Pistol belt.

with stud down, double belt back through sliding keeper and secure end hook in the particular center eyelet required to give proper length. Slide magazine pouch over attached female buckle end and attach it to the fastener provided. The belt is put on with male buckle on the man's right.

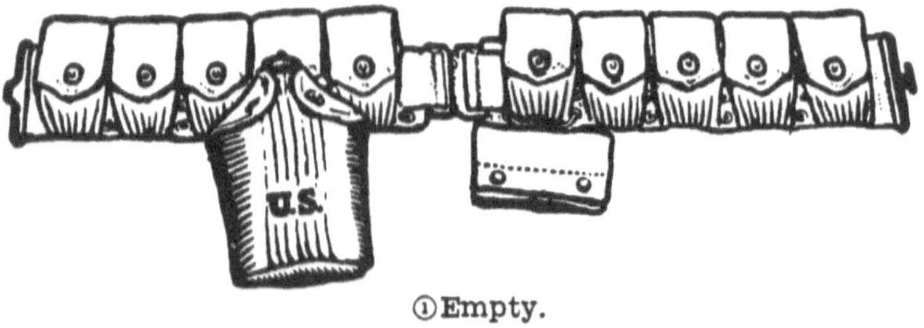

①Empty.

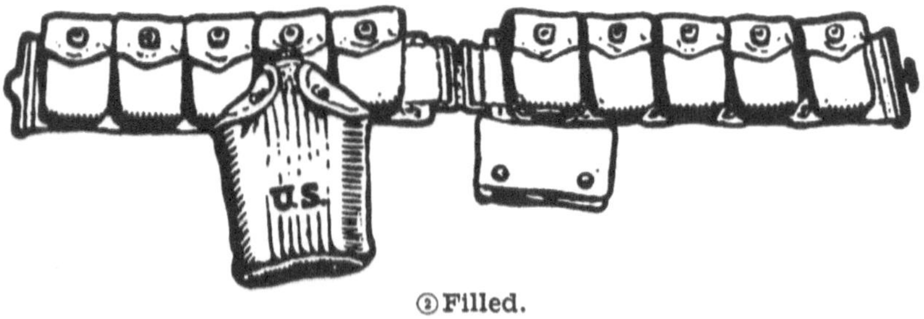

②Filled.

Figure 6.—Belt, cartridge, or magazine.

(2) *Cartridge or magazine belt* (fig. 6).—Place adjusting strap on the ground, eyeleted edge to the front; place pocket sections on the ground in prolongation of adjusting strap,

pockets down, tops of pockets to the front; insert end of adjusting strap in outer loop of metal guide from the upper side, carry it under middle bar and up through inner loop; engage hooks on the end of adjusting strap in eyelets provided on inner surface of belt. The belt is adjusted to fit the individual. It should fit loosely about the waist so that when buckled it may rest well down over hip bones and below the pit of the abdomen. Care should be taken that adjustments are made equally from both ends of adjusting strap so that center eyelet will be in the middle of belt. The proper position of the belt is the same, whether filled or empty.

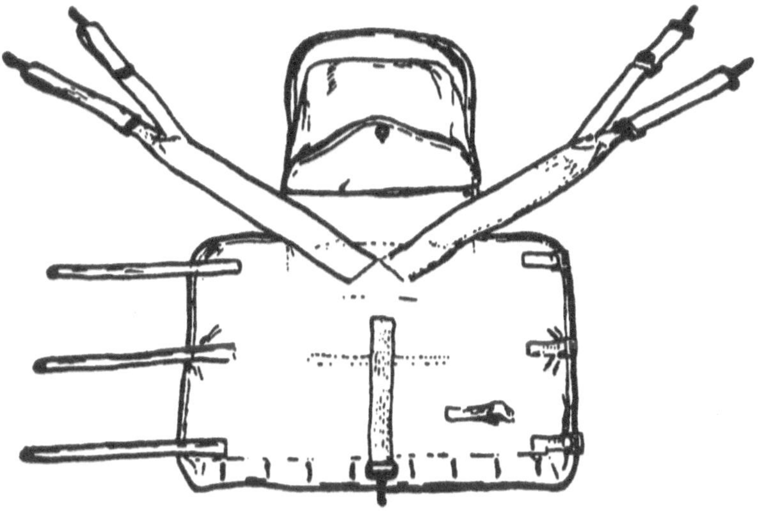

FIGURE 7.—Haversack without pack carrier.

b. To attach pack carrier to haversack (figs. 7 and 8).—Spread haversack on the ground, inner side down, outer flap and meat can pouch to the front. Place button-holed edge of pack carrier, lettered side of pack carrier up, under button-holed edge of haversack. Superimpose button holes of haversack upon corresponding ones of pack carrier. Lace the pack carrier to haversack by passing the ends of coupling strap down through corresponding button holes of haversack and pack carrier nearest the center, bringing the ends up through next button holes and continuing to the right and left, respectively, to the sides.

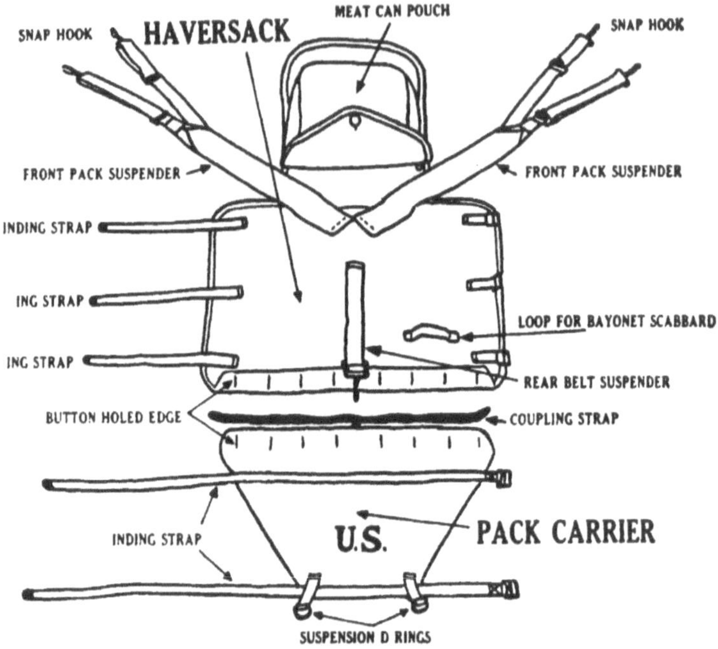

① Before joining (outside view).

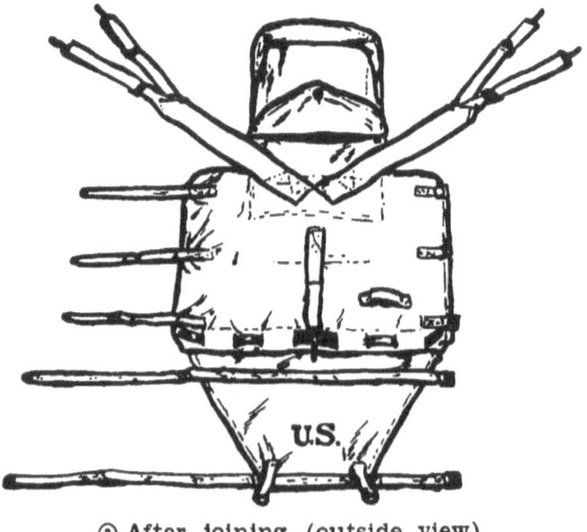

② After joining (outside view).

FIGURE 8.—Haversack and pack carrier.

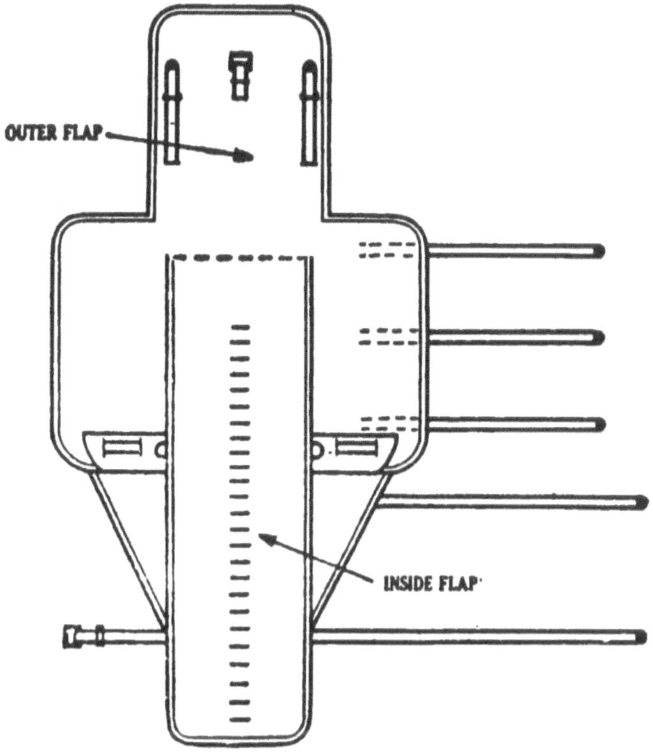

③ Inside view.

FIGURE 8.—Haversack and pack carrier—Continued.

c. To attach haversack.—(1) *To cartridge or magazine belt.*—Place haversack and pack carrier (assembled) on the ground, inner side down; place belt along the junction of haversack and carrier, pockets down, tops toward the haversack; insert hooks on rear belt suspenders in upper rear eyelets of each belt section so that point of hooks will be on the outside of belt; twist front belt suspenders toward the ends of belt and insert snap hooks in eyelets between first and second pockets from each end of belt so that the points of the snap hooks will be on the outside of belt and suspenders will be flat on the body.

(2) *To pistol or revolver belt.*—Proceed as with the cartridge belt, the hooks on the ends of front belt suspenders being inserted in the third eyelets from the buckle, and hooks

on rear belt suspenders in two of the upper eyelets respectively to right and left of center of belt.

d. To attach bayonet scabbard.—(1) *To haversack.*—Attach scabbard by passing its lower end through loops provided on the side of haversack body; then engage double hook attachment in eyelets on outer flap of haversack, inserting hook from the under side. Place bayonet in scabbard, ring to the rear. (See fig. 14 ③.)

(2) *To cartridge belt.*—When the haversack is not carried, attach scabbard to left side of cartridge belt over left hip. Place bayonet in scabbard, ring to the front.

e. To attach intrenching tool or machete carrier to haversack.—Fold outer flap of haversack over so that meat can pouch is uppermost; pass intrenching tool carrier underneath meat can pouch and engage double hook attachment in eyelets in flap provided, inserting hooks from the under side. Secure intrenching tool to roll by means of third haversack binding strap. (See fig. 14 ③.)

f. To make the roll (fig. 9).—Spread shelter half on the ground and fold in triangular end so that shelter half forms a rectangle. Make a second fold by carrying folded edge to opposite edge. Fold the blanket twice parallel to its longer axis so that blanket is now one-fourth its previous width, and then fold once at the middle so as to bring the ends together. Place blanket symmetrically in center of folded shelter half; place underwear, socks, and handkerchief between folds of blanket. Place tent pole on that end of the blanket from which the rolling is to begin. Place pins as shown in figure 9 ②. (An alternate method is to place pins next to and parallel with pole.) Fold sides and then the near end of shelter half snugly over the blanket; fold 10 inches of far end of shelter half toward the blanket and, beginning at near end, roll tightly into folded end of shelter half, thus making an envelope roll.

g. To pack the haversack (fig. 10).—(1) Place equipment on the ground, inside of haversack up, pockets of belt up, haversack spread out, inside flap and pack carrier extended to their full length. Place rations in the center of haversack in front of and touching line of attachment of inside flap. Place toilet articles in front of rations. Fold inside flap of haver-

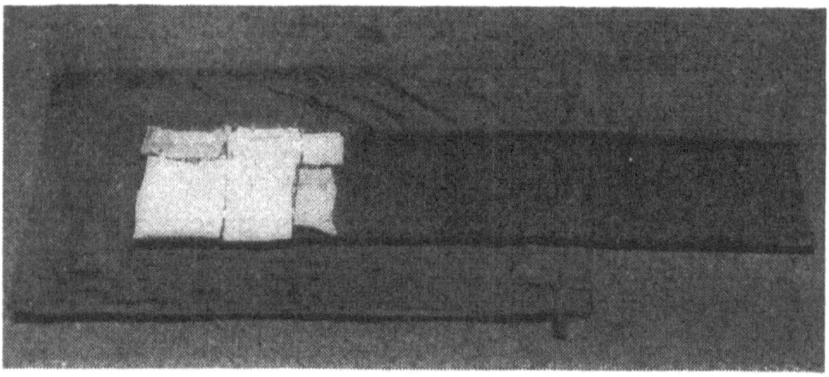

① First step.

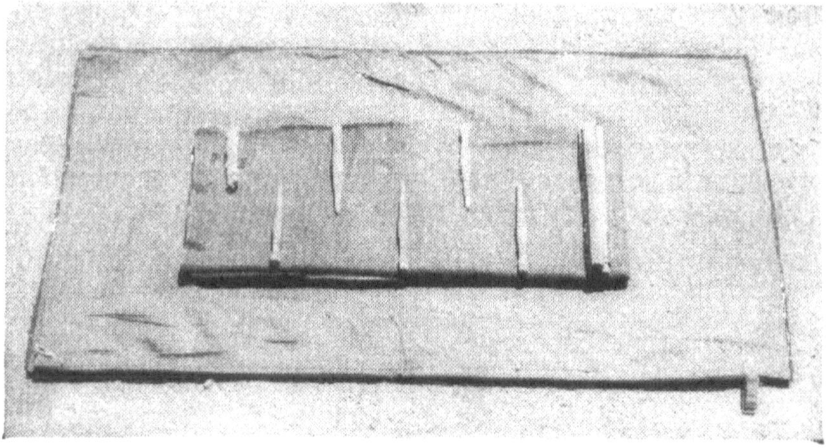

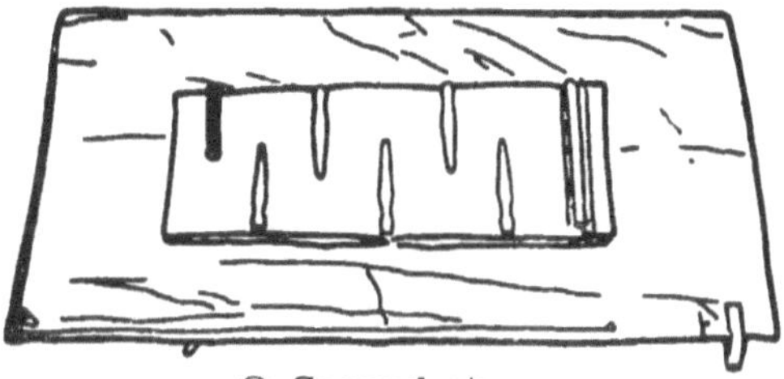

② Second step.

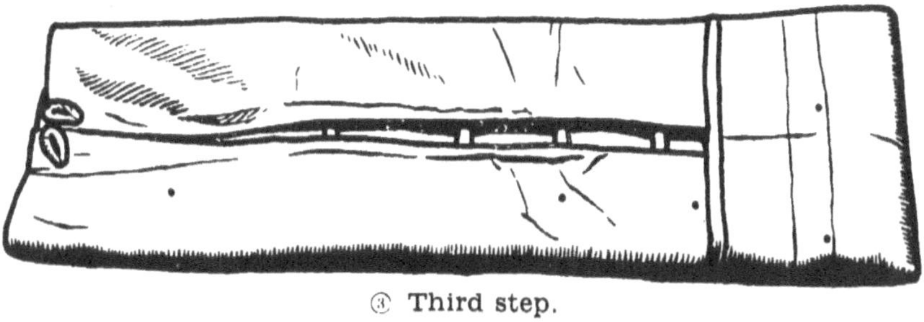

③ Third step.

FIGURE 9.—To make the roll.

④ Completed roll.

FIGURE 9. To make the roll—Continued.

sack over these articles. Fold sides of haversack over rations and toilet articles. Pass upper two binding straps through loop opposite point of attachment of the strap to haversack body. Fasten each strap by passing end of strap through the opening of its opposite buckle next to the buckle attachment, over center bar and back through opening of buckle away from attachment. Pull strap tight and make fastening secure. Fold over outer flap of haversack and fasten it by means of the lower haversack binding strap passed through the buckle on inside of outer flap. Pull strap towards the right, drawing outer flap snugly over filled haversack. The haversack is now packed and the carrier is ready for reception of the roll.

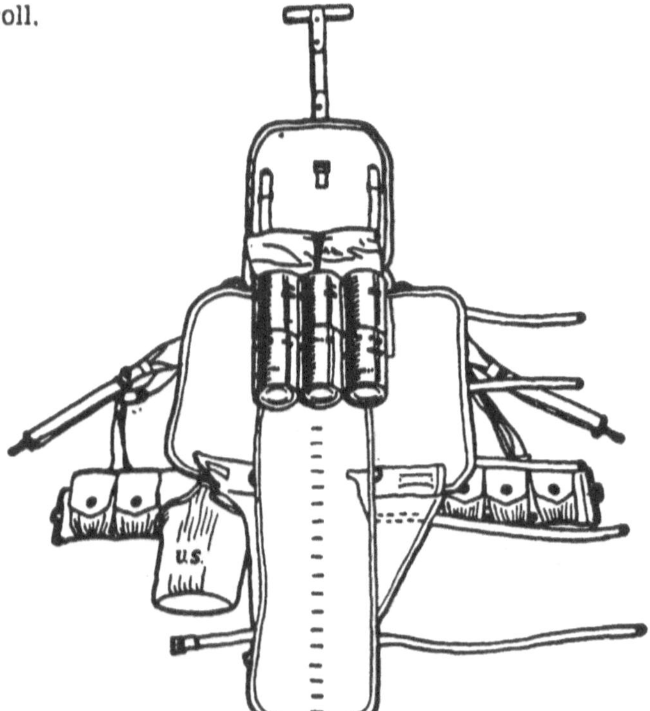

① First step.

FIGURE 10.—To pack the haversack.

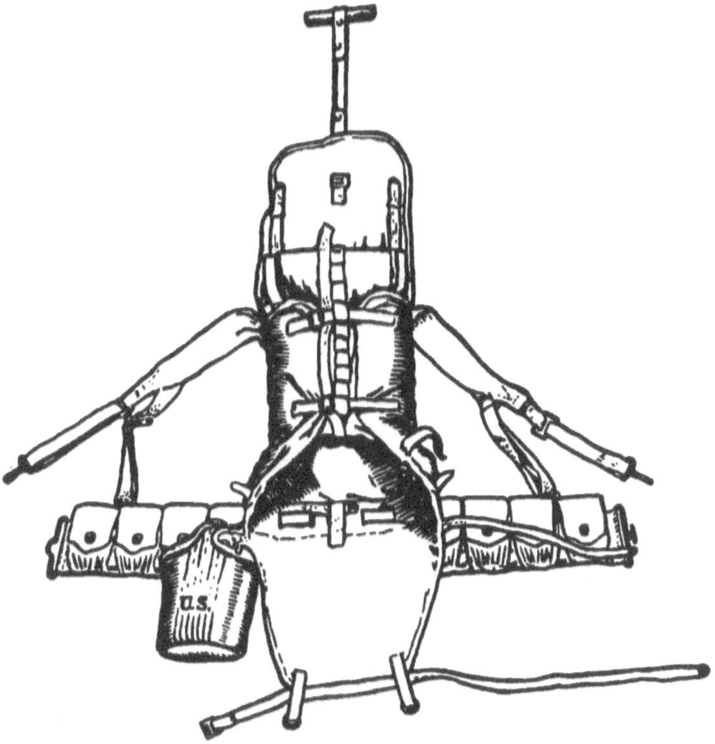

② Second step.

FIGURE 10. To pack the haversack—Continued.

(2) When rations are not carried, roll toilet articles in inside flap so that top of toilet articles will be on line with top of haversack body. Then fold up lower haversack strap against the roll thus formed.

h. To assemble the pack (fig. 11).—Place roll in pack carrier and haversack with one end against bottom of packed haversack. Grasp lower suspension rings, one in each hand. Place right knee against bottom of roll. Pull carrier down and force roll up close against bottom of packed haversack. Without removing the knee, fasten lower pack carrier binding strap over the roll and secure it by passing, from below, its end up through the opening of its corresponding buckle away from the buckle attachment, then over center bar and down through opening of the buckle next to buckle attachment, then back underneath the standing end of strap. In a similar manner secure middle haversack binding strap and then upper carrier binding strap. Engage snap hooks on pack suspenders in lower suspension rings. In order to obtain the maximum benefit from the shoulder loops in suspending the pack on the wearer, each pack should be so assembled that when the roll is carried the length of the assembled haversack and pack carrier is at least 27 inches. This may be done by placing clothing or equipment ordinarily carried

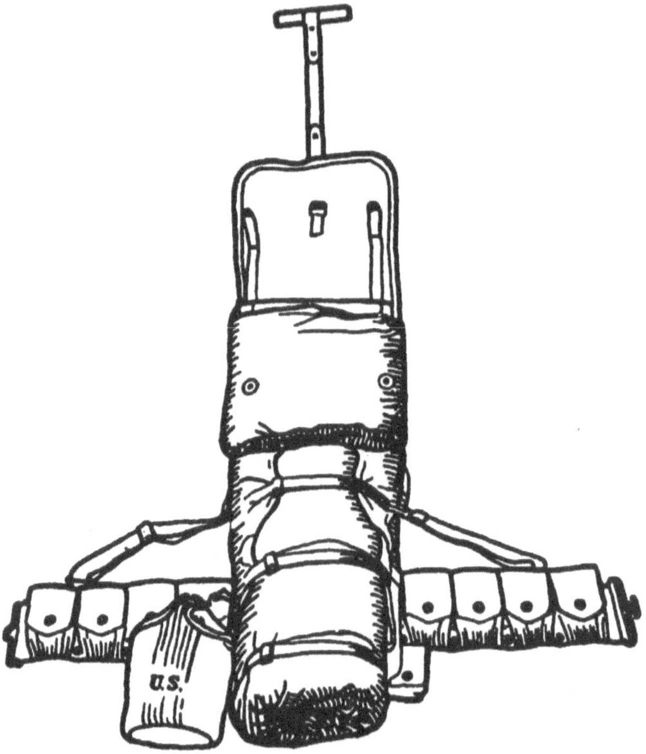

FIGURE 11.—Pack assembled.

elsewhere in the ration space in the haversack so that the assembled pack will be the same length as when rations are carried. When this method is used, the pack carrier will not be folded under.

i. To fold and attach the overcoat (figs. 12 and 13).—Turn sleeves inside out, place overcoat on the ground, outside down, coat smooth, collar extended, sleeves smooth and extended toward pockets; move inside flap of tail under outside flap about 6 inches and gather the slack in the coat thus caused in one fold along middle seam, tapering toward the collar. Fold bottom of front edges of coat about 12 inches toward center seam, forming an approximate parallelogram, the side of which across the coat will be 42 inches. Fold collar end down about 15 inches at the top and roll smoothly and tightly toward the tail; turn tail up to a depth of about 9 inches, and roll entire coat into this pocket. Place the overcoat thus rolled, open side of roll down, on top of haversack, in rear of bayonet handle so that center back seam is over center of top of haversack. Secure coat at the top with a shelter half rope. Bind the ends of overcoat down and along the sides of haversack. Lash the ends of roll to haversack with the rope, using a half hitch near each end of overcoat and passing the rope around haversack over outside flap.

j. To attach the raincoat.—Fold raincoat neatly into a rectangle about 10½ inches long by 8½ inches wide, with outside of raincoat out. Place raincoat between inner and outer

① First step.

FIGURE 12.—To fold the overcoat.

② Second step.

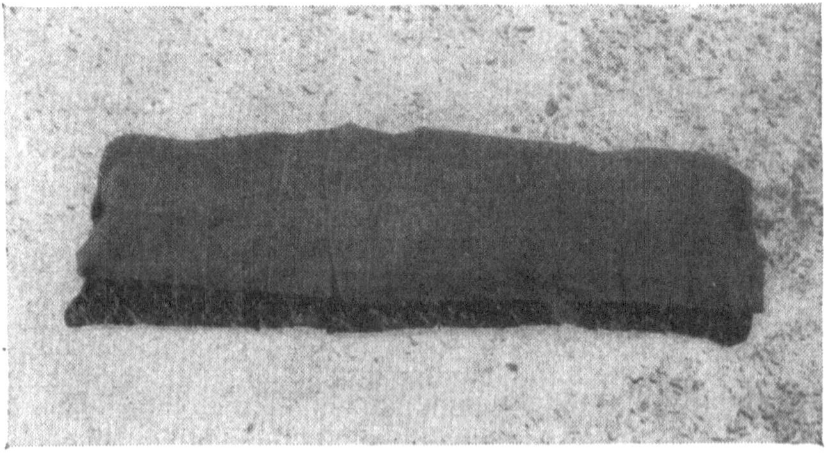

③ Completed fold.

FIGURE 12.—To fold the overcoat.—Continued.

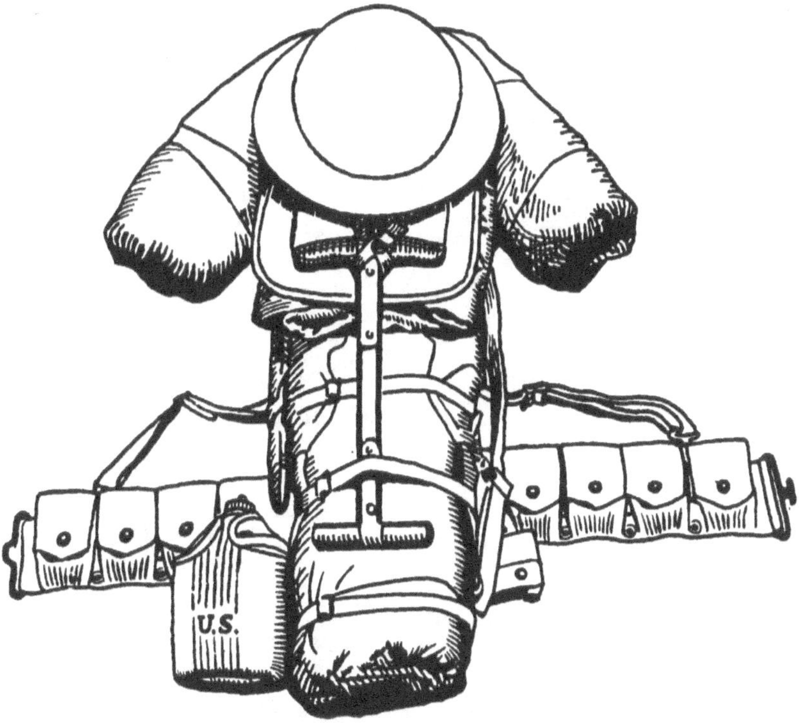

FIGURE 13.—Full equipment with overcoat and raincoat.

haversack flaps. Secure it with lower haversack binding strap by passing the latter under bottom haversack binding strap and fastening it tightly to the buckle on under side of outer haversack flap.

k. To attach the helmet.—Attach and secure the helmet by placing chin strap over meat can pouch.

l. To adjust full equipment.—Put on equipment, slipping the arms through pack suspenders as through sleeves of a coat. By means of adjusting buckles on belt suspenders, raise or lower the belt until it rests well down over hip bones and below pit of abdomen. Raise or lower it in rear until adjusting strap lies smoothly across small of the back. By means of adjusting straps on pack suspenders, raise or lower the load on the back until the top of haversack is on a level with top of shoulders, so that pack suspenders from their point of attachment on the haversack to the shoulders will be horizontal. The latter is essential to proper adjustment of the load. (See fig. 14.)

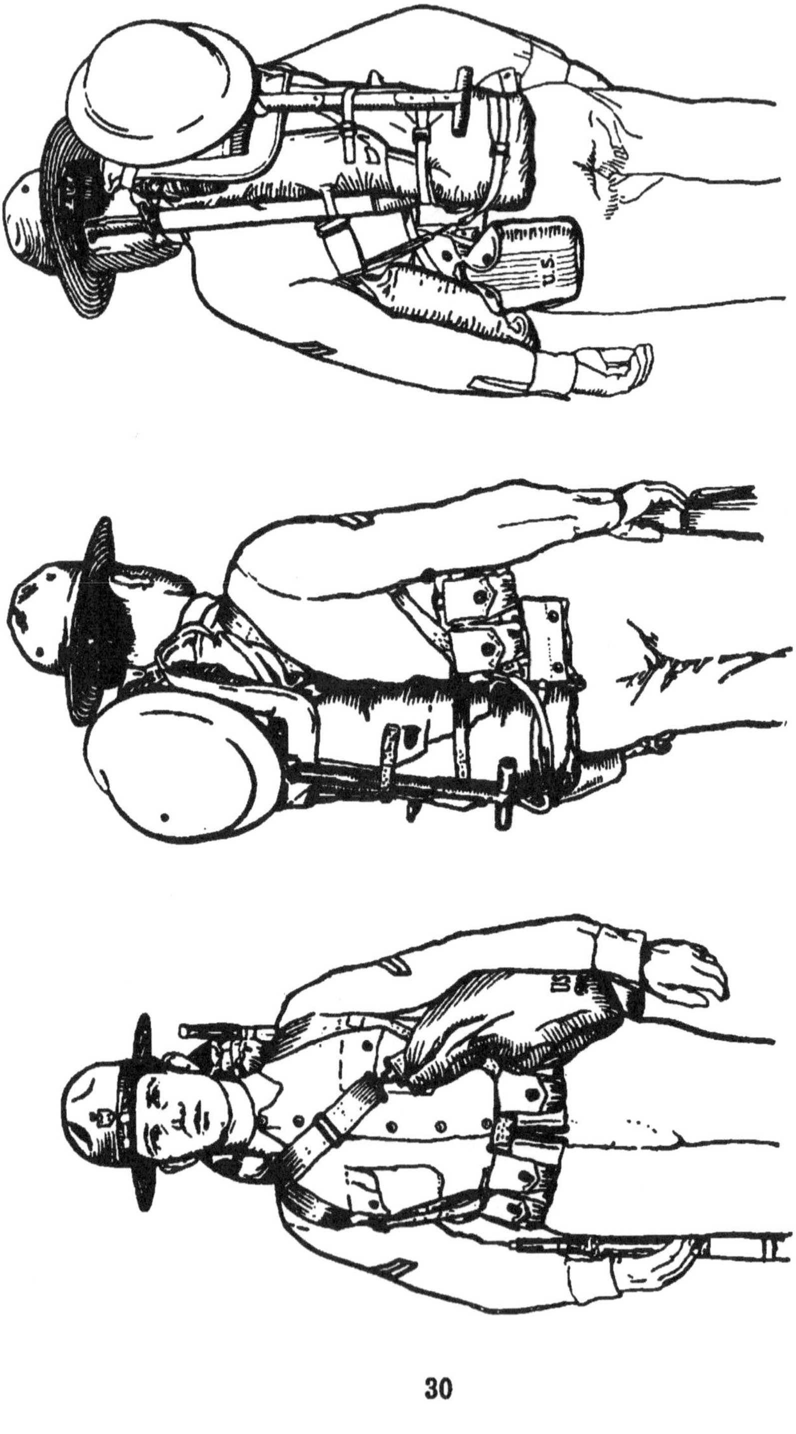

① Front view. ② Right rear view. ③ Left rear view.

FIGURE 14.—Full equipment (less overcoat).

m. To discard the roll without removing equipment from the body.—Unsnap pack suspenders from suspension rings and snap them into eyelets on top of belt and in rear of rear pockets of right and left pocket sections. Support bottom of pack with left hand, with right hand grasp coupling strap at its middle and withdraw first one side and then the other. Pull down on the roll with both hands and remove it. When the roll has been removed, lace coupling strap through buttonholes along upper edge of carrier.

■ 17. METHOD OF PACKING INDIVIDUAL EQUIPMENT ON HORSE.—*a. General.*—(1) The method of packing the saddle of the mounted enlisted man has a decided bearing on the mobility of a command. To maintain this mobility and to prevent sore backs, care must be exercised to insure even distribution of weight on the horse's back. The pressure due to an uneven distribution of arms and equipment on the horse may cause injury to its withers and back. Therefore, weight of articles attached to *one side* of the saddle should be *balanced* as nearly as possible by weight of articles attached to the other side. Weight is better borne by the pommel than the cantle.

(2) The saddle pack of specialists varies from the prescribed basic saddle pack only to the extent required by differences in the equipment carried by each individual on his riding horse.

b. To assemble and pack the cantle or blanket roll.—(1) The shelter tent half is spread flat on the ground, buttons up. The triangular flap is folded over shelter tent half, making the latter a rectangle. The blanket is folded once through the center, parallel to short side, and again through the center perpendicular to short sides. The blanket is then laid on shelter tent half, the longer folded edge parallel to and one inch from long side of shelter tent half, opposite the buttons, the shorter folded edge toward triangular flap, the blanket equidistant from ends of shelter tent half.

(2) The tent pole, folded, is inserted in double fold of blanket, end of pole flush with shorter folded edges, the pole parallel to and fitting snugly into the double fold. The tent pins are inserted in double fold of blanket, near loose edges, placed alternately head and point and overlapping each other so as to occupy about the same space as the tent pole, the pins

parallel to and fitting snugly into double fold of blanket. This leaves a "break" at center of completed roll which allows it to fit the saddle.

(3) The free ends of shelter tent half are folded over corresponding portions of the blanket, the ends of shelter half throughout their length being parallel to its center line. As a prevention against the ends of completed roll pulling out, the free (loop) end of tent rope is passed several times through tent pin loops on the two opposite corners of button side of shelter tent half, stretched flat and tied with a single bow-knot. The button side of roll is folded back about 6 inches (as far as second button) in order to form a pocket when roll is completed, edge of pocket being parallel to edge of blanket.

(4) Using the hands and knees the blanket and shelter tent half are rolled tightly from side opposite buttons into pocket at button side, making a tightly bound roll. The roll is then "broken" or curved to fit the cantle of saddle, with free edge of pocket uppermost and to the rear, so that this edge fits snugly against the roll and prevents entrance of rain or snow when roll is on the saddle. It is advisable for two men to work together in making up the roll.

c. To fold the raincoat and/or overcoat.—The raincoat, inside out, with collar extended is folded once lengthwise. It is rolled tightly from folded edge toward buttons, making length of roll the same as full length of garment. If the overcoat is carried, it is rolled in the same manner as the raincoat.

d. Feed and grain bag.—The grain bag, with or without grain, is secured inside the feed bag.

e. To pack the saddle (fig. 15).—(1) The saddle, off the horse, is placed on the ground, or otherwise as is most convenient for packing. The saddlebags are placed on the cantle and secured thereto by the attachments provided. The cantle roll is strapped to the cantle by means of the cantle straps which are wrapped three times around the roll. The straps are securely buckled and the loose ends, falling to the front, are tucked under the straps. The buckles should be far enough to the rear to prevent injury or discomfort to the trooper. The ends of the roll are pressed forward and down toward the saddlebags. The two outside straps should pass over the roll well down toward its ends in order to keep the roll curved to the shape of the cantle.

(2) The feed bag and grain bag are strapped on top of the pommel bar (under the raincoat or overcoat), regardless of whether grain is carried. The loose ends of the straps, falling to the rear, are tucked under the straps. The raincoat, and/or overcoat, collar to the left, is placed on top of feed and grain bag and strapped on by means of the pommel straps. The ends of the roll are pressed in toward the horse's shoulders. All buckles should be far enough forward to prevent injury or discomfort to the trooper.

(3) The canteen and cup, in their cover, are snapped to the right cantle ring. The cover is fastened tight to the saddlebag by buckling the two rear saddlebag cover straps over the canteen cover strap where it passes under the bottom of the canteen cover.

(4) Attach helmet to near (left) saddlebag by buckling the three straps of saddlebag flap through the chin strap.

(5) The rifle scabbard is secured to left side of saddle, attached to the pommel ring by the upper strap of scabbard and to cantle ring by lower strap. Both straps are so adjusted that the scabbard will hang at an angle of 30° with the vertical. The rifle is not inserted in the scabbard until after the horse is saddled.

(6) Wherever possible, two men work together in placing the packed saddle on the horse's back. In case only one man is available, it may be found advisable to attach saddle bags, canteen, cup and cover, and rifle scabbard after the horse has been saddled.

f. Drivers (artillery).—(1) *General*.—In order to equalize loads on the two horses, certain articles, as prescribed by Table II may be placed on saddle of the off horse instead of on riding horse.

(2) *To pack feed bags*.—To pack feed bags fill the grain bags; tie mouths securely and place a filled grain bag in each feed bag, mouths of grain bags down. Roll feed bags, securing closed ends by means of web straps at ends of feed bags. The snap ends of the straps are either snapped in the rings on feed bags or tucked under turns in straps. The open ends of the two feed bags are closely secured together to prevent lower ends from rubbing against traces. When grain is not carried, fold empty grain bags and place them inside feed

① Near side.

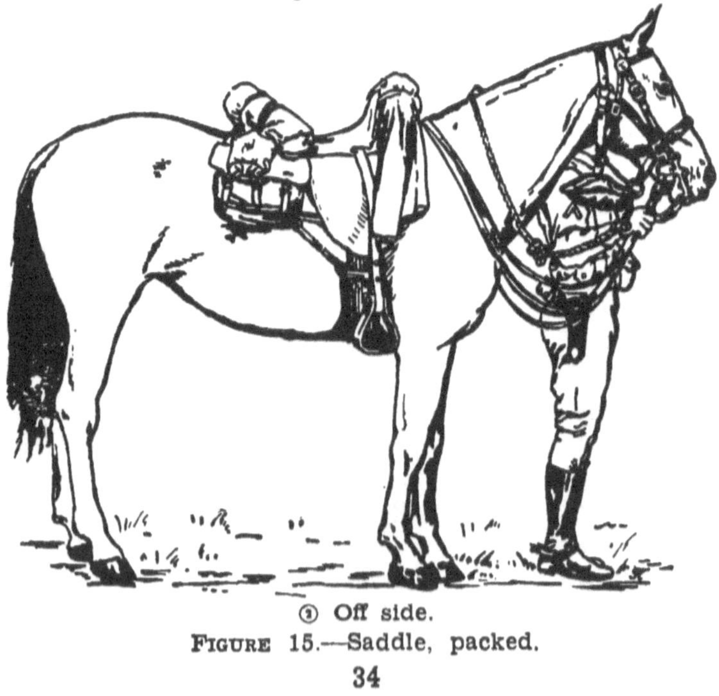

② Off side.

FIGURE 15.—Saddle, packed.

bags. Roll feed bags along their longer edges and secure as above.

(3) *To pack saddle.*—To pack the saddle of the off horse, the near stirrup being passed over seat of saddle, the procedure is as follows:

(*a*) The saddlebags are placed on cantle and secured by attachments provided.

(*b*) Place surcingles under straps of saddlebag flaps above the loops in saddlebag flaps through which the straps pass.

(*c*) Attach mounted canteen cover to left rear cantle ring by the snap of canteen strap.

(*d*) Attach helmet to off (right) saddlebag by buckling the three straps of saddlebag flap through the chin strap.

(*e*) Extend center cantle strap. Place roll on top of buckle end of cantle strap in the center of seat of saddle so that open edge will be down. Place the two feed bags, secured together at their open ends, across seat of saddle in front of roll. Bring tongue end of cantle strap to the front over center of roll and junction of the two feed bags; take one turn around the junction of the two feed bags; pass strap to rear under roll, bring it forward over center of roll, and buckle it. Each coat strap is passed under rear quarter strap and once around feed bag about 4 inches from the lower end, punching a throat in feed bag to prevent strap from slipping; pass each strap to rear and once around roll about 6 inches from end of roll; bring strap from under roll, over itself at the interval between feed bags, roll and buckle. The ends of roll are drawn close to ends of feed bags before being secured. The coat straps are attached to saddle bag side—strap rings about saddlebag side straps to avoid twisting the rings.

(*f*) Place overcoat, collar to the left, across seat of saddle in front of feed bags and secure by two pommel coat straps. When raincoat and overcoat are both carried, place raincoat on top of overcoat. When overcoat is not carried, raincoat is packed as provided for overcoat. (See fig. 16.)

(4) *To pack saddle of the off horse when blanket rolls and feed bags carried on limber.*—To pack the saddle of the off horse, the near stirrup being passed over seat of saddle, the procedure is as follows:

(*a*) Place the saddlebags on the seat of the saddle and secure them by passing the saddlebag straps through the cinch rings and drawing them tight.

(*b*) Place the overcoat on top of the saddlebags, collar to the left; place the raincoat on top of the overcoat. Take a couple of turns with the middle cantle strap around the two coats. Then take several turns around each end of the coats, using a coat strap on each end. Place the free end of each strap through the cinch ring on each side and pull the coats down firmly against the saddle.

(*c*) Secure end of halter tie rope of off horse to off pommel ring of off saddle, and that of near horse to near pommel ring of near saddle.

FIGURE 16.—Driver's (artillery) off horse.

■ 18. METHOD OF ATTACHING EQUIPMENT TO VEHICLES.—*a. To limber.*—Extend the four blanket straps and place the blanket rolls across the top of ammunition chest in rear of bucket brackets, feed bags on top of blanket rolls; then buckle blanket straps.

b. To motor vehicles.—When carried on tanks, scout cars, tractors, trucks, motorcycles, and passenger cars, the disposition of rolls will depend upon the particular construction of the vehicle. Vehicles of strictly commercial type are not provided with straps or other fastenings for securing the rolls. It is then necessary to utilize the straps pertaining to the rolls themselves or to use ropes or special straps. It will often be preferable to dispose the rolls within the vehicles, thus reducing danger of loss and affording better protection from weather.

SECTION IV

DISPLAY OF BASIC EQUIPMENT FOR FORMAL INSPECTION

■ 19. ENLISTED MAN WITH HAVERSACK AND PACK CARRIER OR SIMILAR EQUIPMENT.—*a*. When shelter tents are not pitched, the relative position of basic articles of equipment is as indicated in figure 17.

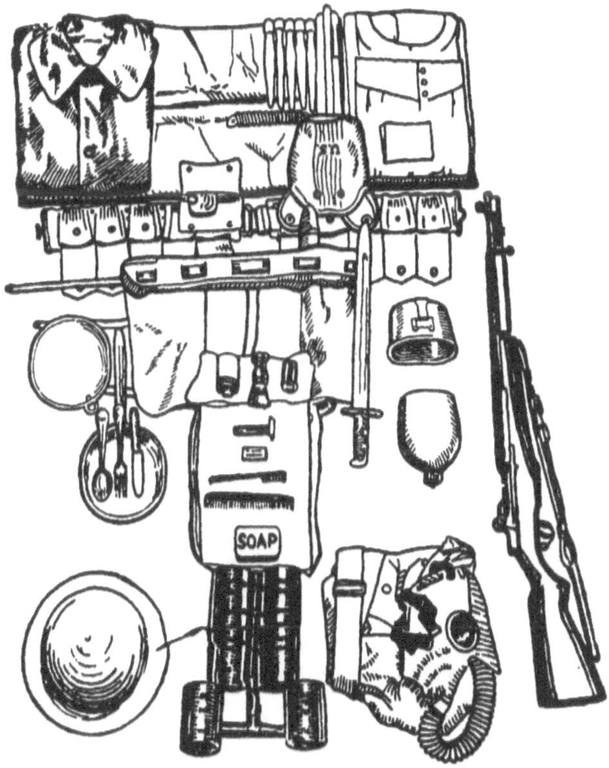

FIGURE 17.—Display of basic equipment of enlisted men (dismounted).

b. Special items of equipment not shown on figure 17 are *added* to the display without disturbing the relative positions of basic articles.

c. When shelter tents are pitched, the blanket remains in the position shown for shelter half and blanket in figure 17. Rear edge of blanket is on line with front (entrance) edges of shelter tents as shown in figure 20.

■ 20. MOUNTED ENLISTED MEN OR THOSE SIMILARLY EQUIPPED.—*a*. When shelter tents are not pitched, the relative position of basic articles of equipment is as indicated in figure 18.

b. Special items of equipment not shown on figure 18 are *added* to the display without disturbing the relative positions of basic articles.

c. When shelter tents are pitched, the blanket remains in the position shown for shelter half and blanket in figure 18. Rear edge of blanket is on line with front (entrance) edges of shelter tents as shown in figure 20.

SECTION V

TO PITCH SHELTER TENTS

■ 21. GENERAL.—The purpose of this instruction is to enable the unit commander to establish his unit expeditiously in shelter tents. Except at inspections, alinements and intervals of the tents will be subordinated to the employment of the available ground for the concealment and comfort of the men.

■ 22. FORMATIONS.—*a*. The company (battery or troop) commander indicates the area for each platoon (or section) or the line on which its tents are to be pitched.

b. For instruction and for inspections, the platoons or sections are formed for pitching tents as prescribed in FM 22–5.

■ 23. FORMAL SHELTER TENT PITCHING.—*a*. When directed by platoon leader (or chief of section), odd numbered men mark their positions by thrusting bayonets, or by marking with their left heel, points alongside the outside of their left heels, near their insteps. The locations marked indicate the positions of the front tent poles. Odd and even numbers (numbers one and two; numbers three and four; etc.) pitch tents together.

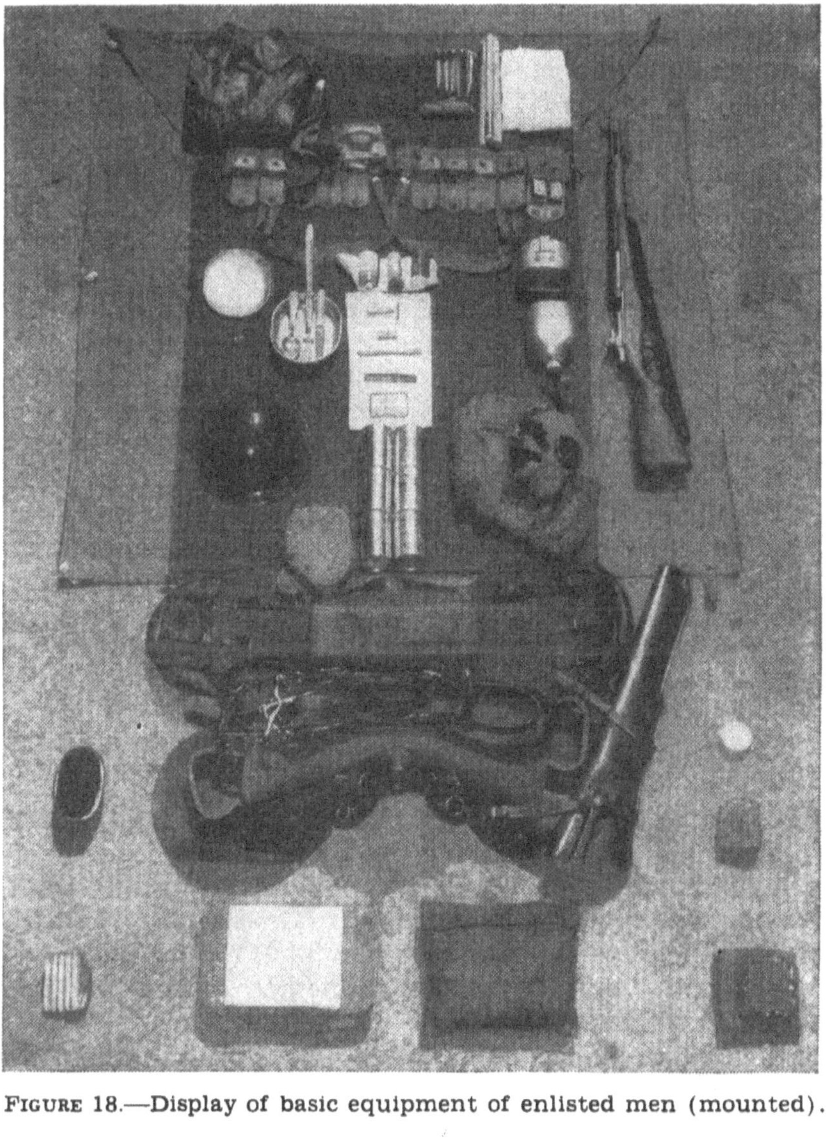

FIGURE 18.—Display of basic equipment of enlisted men (mounted).

b. At the command, PITCH TENTS, each man (if armed with the rifle) steps off obliquely with the right foot a full pace to the right front, lays his rifle on the ground, muzzle to the front, barrel to the left, butt near the toe of his right foot. He then steps back into place. All men then unsling equipment and place their packs (or rolls) on the ground in front of them, haversacks (saddle bags or field canvas bags) up and to the front, the packs two paces in front of their positions. They then open their packs and remove their shelter halves, poles, and pins. Each odd numbered man, when not armed with a bayonet, places a pin in the ground on the spot which he previously marked with his left heel. The men of each pair spread their shelter halves on the ground which the tent is to occupy, triangle to the rear, buttons to the center, the even numbered man's half on the left.

c. They then button the halves together. The odd numbered man adjusts his pole through the eyelets in the front of the tent and holds the pole upright in position beside the bayonet (or pin). The even numbered man pins down the front corners of the tent in line with the bayonets (or pins). He then drives the front guy pin a rifle length in front of the front pole. If he is not armed with the rifle, he measures this distance with his tent rope by taking the distance from the base of the front tent pole to one of the front tent pins. He places the loop of the guy rope over the front guy pin, runs the other end of the rope through the loops of the shelter halves and ties it, making sure that the pole is vertical when the rope is taut. The even numbered man then adjusts the rear tent pole through the eyelets in the rear of the tent. The odd numbered man pins down the rear of the tent and drives the rear guy pin so that it is a bayonet length in rear of the rear pin of the triangle. If he is not armed with the bayonet, he drives the rear guy pin two and a half tent pin lengths from the rear triangle pin. He then adjusts the guy rope. The even numbered man then drives the remaining pins on the left of the shelter tent and the odd numbered man drives them on the right.

d. Each man displays the contents of his pack as described in paragraphs 19 and 20. After the equipment of all men is displayed, the platoon (or section) reforms in single rank with normal interval in front of their equipment, the right flank man taking position about one foot in front of the front guy pin of his tent.

FIGURE 19.—Pitching shelter tents.

FIGURE 20.—To display equipment.

■ 24. **Pitching Double Shelter Tents.**—*a.* The double tent camp is preferable to the single tent camp in cold or inclement weather because of the added shelter it affords. It is useful in restricted areas because of its greater concentration of personnel. The double tent is composed of two single tents buttoned together at the square *ends*. The tent is supported by three poles, one in the center and one at each end.

b. The platoons (or sections) are formed for pitching tents as prescribed in FM 22–5, except that, after all squads are in line, the platoon (or section) is formed at close interval.

c. On direction of the platoon leader (or chief of section) number one and every fourth man thereafter mark the spot alongside the outside of their left heels, near their insteps. This indicates the positions of the front tent poles. Numbers one, two, three, and four pitch and occupy one double tent. Numbers five, six, seven, and eight the next double tent, and so on.

d. At the command, PITCH DOUBLE SHELTER TENTS, each man (if armed with the rifle) steps off obliquely with the right foot a full pace to the right front, lays his rifle on the ground, muzzle to the front, barrel to the left, butt near the toe of his right foot. He then steps back into place. All men then unsling equipment and place their packs on the ground in front of them, haversacks (saddle bags or canvas field bags) up and to the front, the packs two paces in front of their positions. They then open their packs and remove their shelter halves, poles, and pins. Each odd numbered man, when not armed with a bayonet, places a pin in the ground on the spot which he previously marked with his left heel. The four men of each group pitching tents together spread their shelter halves on the ground which the tent is to occupy as follows: the lower odd numbered man at the right front, the lower even numbered man at the left front, the higher odd-numbered man at the right rear and the higher even numbered man at the left rear.

e. They then button together the four shelter halves of each tent, the ridge first, then the square ends so that the center eyelets of the shelter halves will be in the following order; the lower half of the front tent on the bottom, the lower half of

FIGURE 21.—Laying out shelter halves for double shelter tents.

the rear tent next, the upper half of the front tent next, and the upper half of the rear tent on the top. The odd numbers insert and support the front and rear poles, respectively. the front pole in the location marked, the rear pole in a line with it perpendicular to the line of front poles. The even numbers pin the front and rear corners of the tent. One even number then enters the tent with his pole and, assisted by the other, inserts the pole through the center eyelets of the shelter halves.

f. The even numbers then drive two guy pins at each end of the tent. Guy pins are one rifle's length distance from the corner pins of the tent and on line with the side pins. They then adjust the guy ropes through both loops at once at each end of the tent. The even numbers then fasten the guy ropes and all drive the remaining pins. The occupants turn back the triangular ends, and, for inspections, display the contents of their packs as prescribed in section IV. Each man stands at attention in front of his own shelter half, toes on line with the guy pins. If equipment is not to be displayed the platoon (or section) reforms in single rank with close interval as shown in Figure 22, the right flank man taking position so that his left foot is on line with the tent poles of his tent.

■ 25. To Strike Shelter Tents.—The men are directed to fall in at their own tents. They take the positions as shown in figure 20 or figure 22. At the command, STRIKE TENTS, the men pull the pins, unbutton the shelter halves, and roll their packs. When packs are rolled men equipped with haversacks place their packs on the ground in front of them, haversacks to the front, suspenders up, packs one foot in front of the toes, and stand at ease in the positions they occupied before the command to pitch tents. Equipment is not slung until directed by the platoon leader (or chief of section).

Section VI

TO PITCH WALL AND PYRAMIDAL TENTS

■ 26. Wall Tents.—*a. Pitching.*—(1) *Preliminary arrangements.*—The officer or noncommissioned officer in charge selects suitable ground. He indicates the direction in which the tent is to face, the line on which the tent is to be placed,

FIGURE 22.—Double shelter tents pitched with men at attention. (Equipment not displayed.)

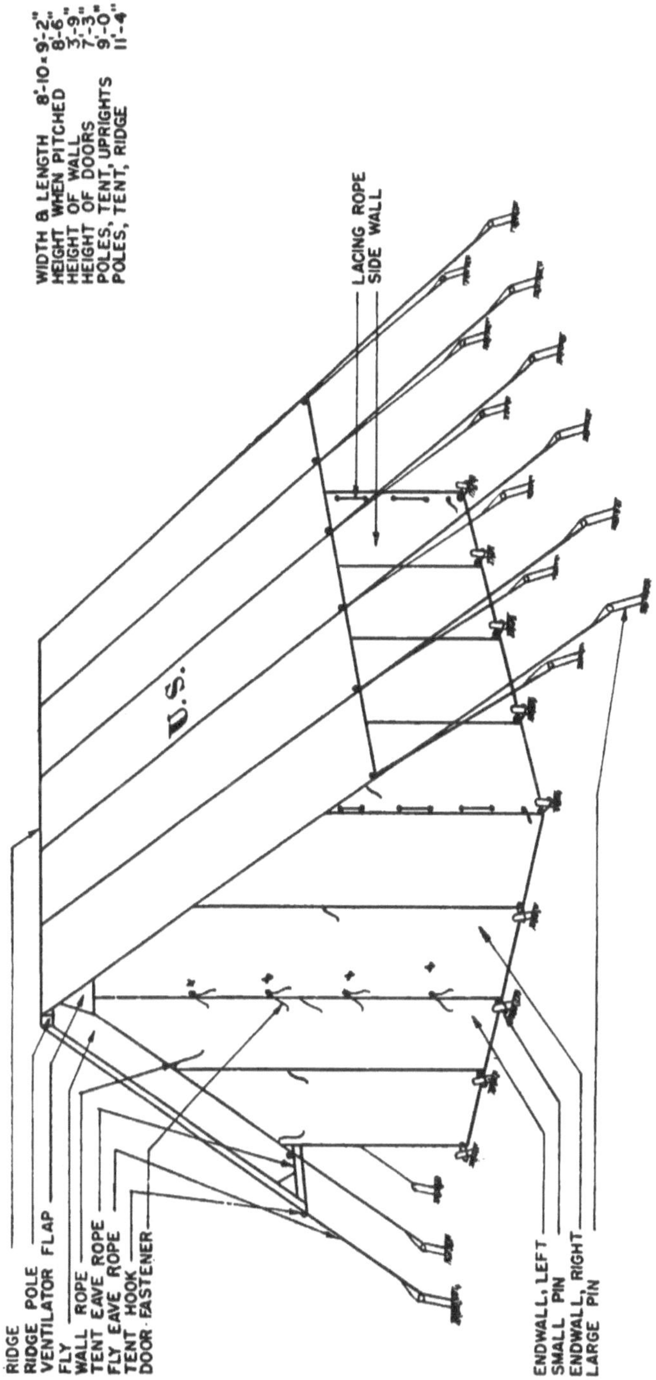

FIGURE 23.—Wall tent.

and the position of the door pin. Four men are required to pitch each tent.

(2) *Procedure for each tent.*—(*a*) First, a pin is driven to mark the center of the door. Then the tent is spread on the ground it is to occupy, the door loops being placed over the door pin. The front corners are drawn taut, alined, and pegged down. The rear door is laced, if there is one. Rear corners are drawn taut in both directions and pegged down. Then the four corner guy pins are driven in prolongation of the diagonals of the tent and about two paces beyond the corner pins.

(*b*) The front door is temporarily loosened and the two corner loops on one side only are loosened from the corner pins. The ridge pole is inserted between the ventilation strip and the tent ridge. The upright eyelet pins are inserted in the ridge pole and in the eyelets of the tent. The fly eyelets are placed over the end-pole eyelet pins. The tent is raised and held in position. The two loosened corner loops are replaced and the corner and fly guy ropes secured and tightened to hold the poles vertical. The wall pins are driven through the open loops as they hang. Finally, the intermediate guy pins are driven in alinement with the corner guy pins already driven.

b. Striking.—The men first remove all pins except those of the four corner guy ropes. The pins are neatly placed in a receptacle. One man holds each corner guy rope, and the tent is lowered. The poles are removed and the tent folded. The poles are fastened together and the remaining pins collected.

c. Folding.—(1) First the tent is spread flat on the ground, folded at the ridge so that the bottoms of the side walls are even, ends of the tent forming triangles to the right and left. Then the triangular ends of the tent are folded in toward the middle to form a rectangle. The top is folded over about 9 inches. The tent is folded in two by carrying the top fold over to the foot, and again from the top to the foot.

(2) All the guy ropes are thrown on the tent except the second from each end. The ends are folded in so as to cover about two-thirds of the second widths. The left end is then folded over to meet the turned-in edge of the top right end,

then the right end is folded over the top right end, next the right end is folded over the top, completing the bundle. Finally, the bundle is tied with the two exposed guy ropes.

■ 27. PYRAMIDAL TENTS.—*a. Pitching.*—(1) *Preliminary arrangements.*—The commander designates the ground on which the tents are to be erected. The line of tents is marked by driving a wall pin on the spot to be occupied by the right front corner of each tent. The interval between adjacent marking pins should be 30 feet, which leaves a space of about 2 feet between tents. Each tent is usually erected by the squad which will occupy it.

(2) *Procedure for each tent.*—(*a*) If the tripod is to be used, it is first spread on the ground where the center of the tent will be.

(*b*) The tent is next spread on the ground which it is to occupy, door to the front. The corners and door are laced and the right front wall loop is placed over the corner pin already driven.

(*c*) The left corner wall loop is carried to the left as far as it will go, and a wall pin driven through it, in line with the right corner pins already driven.

(*d*) The rear corner wall loops are pulled to the rear and outward, so that the bottom of the rear wall of the tent is stretched to complete the square. Wall pins are then driven through these loops. Each rear corner pin should be directly in rear of its corresponding front corner pin to form the square. Unless the canvas is wet a small amount of slack should be allowed before the corner pins are driven.

(*e*) Three men crawl under the tent and fit the pole into the rings, and such accessories as hood, fly, and tripod are adjusted.

(*f*) Steadied by one man at each corner guy rope, the tent is then raised by the men underneath.

(*g*) The four corner guy ropes are then placed over the lower notches of large pins, which are driven in prolongation of the diagonals at such distances as to hold the walls and ends of the tent vertical and smooth when the guy ropes are drawn taut.

(*h*) A wall pin is then driven through each remaining loop, and a large pin for each guy rope is driven in line with

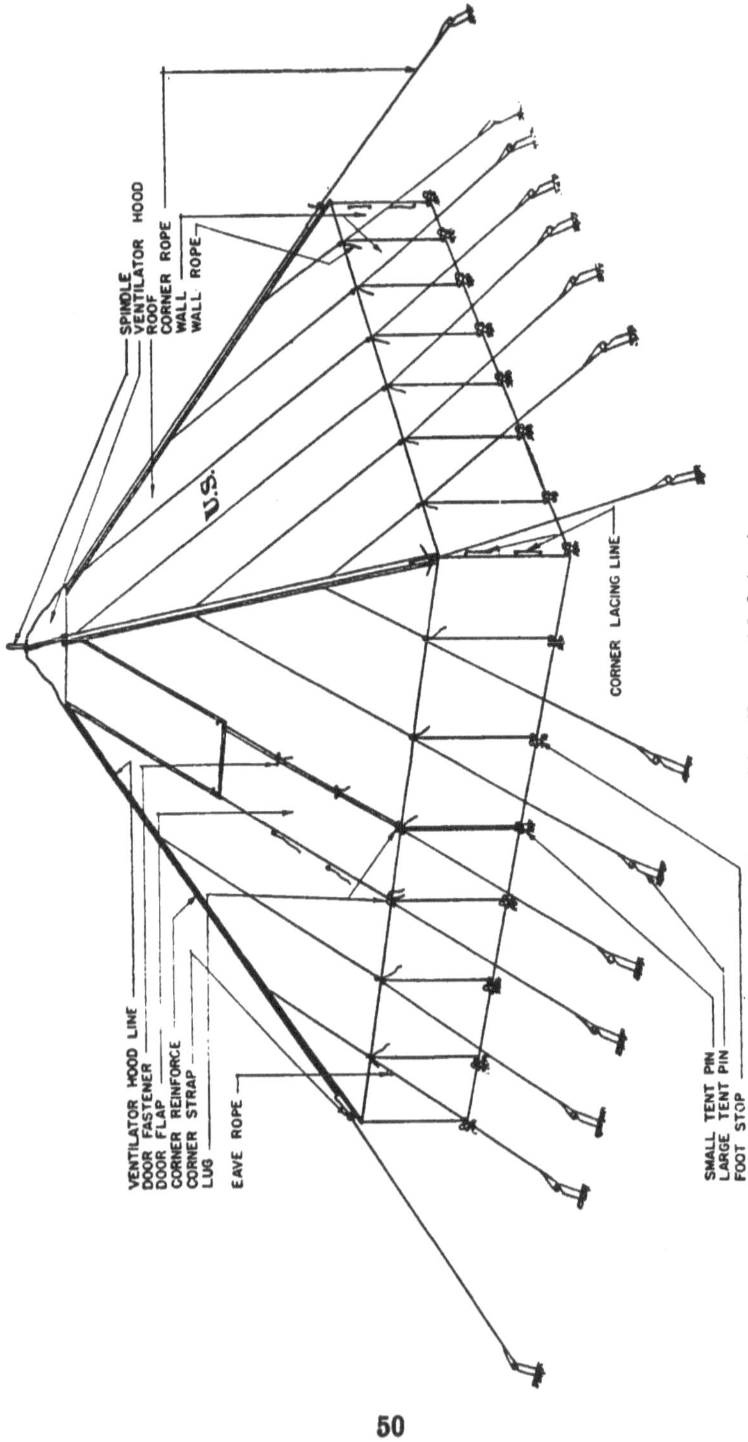

FIGURE 24.—Pyramidal tent.

the four corner guy pins already driven. The guy ropes of the tent are placed over the lower notches of the large pins. All ropes are then drawn taut.

b. Striking.—First, all pins are removed except those of the four corner guy ropes and the rear corner wall pins. The pins are neatly piled or placed in a receptacle. With one man holding each corner guy rope, the tent is then slowly lowered to the rear. The tripod and poles are fastened together and the remaining pins collected.

c. Folding.—(1) *Procedure for each tent.*—(*a*) The tent having been lowered to the rear, the back wall and roof canvas are pulled out smooth. This is accomplished by leaving the rear corner wall pins in the ground with the loops attached. One man at each corner guy and one or two holding the square iron perpendicular pull the canvas to its limit away from the former front of the tent. This places the three remaining sides of the tent on top of the rear side, with the door side in the middle.

(*b*) Next, to straighten the right side wall and roof canvas, the right front corner is carried over and laid on the left front corner. All canvas is pulled smooth, the guy ropes thrown toward the square iron, and the bottom edges pulled even. Then the right front corner is returned to the right to cover the right rear corner. This folds the right side of the tent on itself, with a crease in the middle and under the front side of the tent.

(*c*) Next, to straighten the left side wall and roof canvas, the left front corner is carried to the right and back in a similar fashion. This will leave the front and rear sides of the tent lying smooth and flat and the two side walls folded inward, each on itself.

(*d*) The square iron is folded downward toward the bottom of the tent and then the hood is placed in the square iron. The tent is now folded around the square iron as a core, all folds being pressed down flat and smooth, and parallel to the bottom of the tent. If each fold is compactly made and the canvas is kept smooth, the last fold will exactly cover the lower edge of the canvas.

(*e*) All exposed guys are laid along the folded canvas except the two on the center width. These two are pulled out and away from the bottom edge to their extreme length so as to be used later for the final tying of the bundle.

(*f*) From one end the bundle is folded toward the center hinging on the first seam; that is, the one joining the first and second widths. It is again folded toward the center, so that the canvas already folded will come to within about 3 inches of the middle width. It is again folded to the opposite edge of the middle width of canvas.

(*g*) From the opposite end the first width is folded in half. A second fold comes about 4 or 5 inches from the part of the tent already folded from the first end. This second fold is then thrown entirely over the part already folded.

(*h*) The exposed guys are drawn taut toward and across one another so that they are at right angles. The bundle is turned over on the guy rope which is in prolongation of the tent seam. The guy ropes are crossed on the new top of the bundle. The bundle is turned over again on the crossed ropes and tied.

(2) *Bundle.*—(*a*) When properly tied and pressed together the bundle will be about 11 by 23 by 34 inches.

(*b*) The unit designation, stenciled on the lower half of the middle width of canvas in the back wall, will appear on the exposed top of the bundle.

SECTION VII

TO PITCH LATRINE SCREENS

■ 28. PITCHING.—*a. Preliminary arrangements.*—A detail of six men, supervised by a noncommissioned officer, is required to pitch the latrine screen. The noncommissioned officer selects suitable ground, indicates the direction in which the screen is to face, the line on which it is to be placed, and the position of the inside corner of the entrance.

b. Procedure for each latrine screen.—(1) A tent pin is driven to mark the inside corner of the entrance. The latrine screen is unfolded and the end having both a long and a short tie rope (as differing from the end having two short tie ropes) is placed near the entrance pin. The short tie rope on the bottom of the screen is tied to this pin.

(2) The latrine screen is laid on the ground following generally its final outline (see fig. 25). A long ridge pole and three uprights are assembled and raised to a vertical position near the center of the latrine screen. The uprights being held in position, two center guy rope pins are driven in two paces from the front center upright and generally in prolongation of the diagonals from the right and left rear corners of the latrine screen through the front center upright (see fig. 25 ②). A third guy rope pin is driven about one pace from the rear center upright and on line with the three uprights. Guy ropes are placed over these pins and over the upright eyelet pins and tightened. The latrine screen is raised on the outside of the inner and rear uprights and is tied to the ridge pole with the bottom about six inches from the ground.

(3) At the narrow end of the screen a short ridge pole and two uprights are assembled and held in position. Guy rope pins are driven in about two paces from the uprights and in prolongation of the diagonals from the inner center and rear center uprights. Guy ropes are attached as before. The latrine screen is raised on the outside of the uprights and is tied to the ridge pole so that the bottom is about six inches from the ground.

(4) Before the long ridge pole and uprights are assembled to form the wide end of the latrine screen, the long tie rope from the top of the inside corner of the entrance is passed over the ridge pole. The upright eyelet pins are inserted in the two outside holes of the ridge pole and the uprights are raised and held in place. The two remaining guy rope pins are then driven. All four front guy rope pins are not on the same line. (See fig. 25②.) Guy ropes are attached and tightened. The latrine screen is raised and tied to the end ridge pole and the outside of the entrance is tied to the center ridge pole.

(5) The long tie rope from the top of the inside corner of the entrance is drawn taut. All tie ropes on the lower edge of the latrine screen are tied to the uprights. Any slack is taken out of the screen by a final tightening of the guy ropes and if necessary by shifting slightly the upright poles.

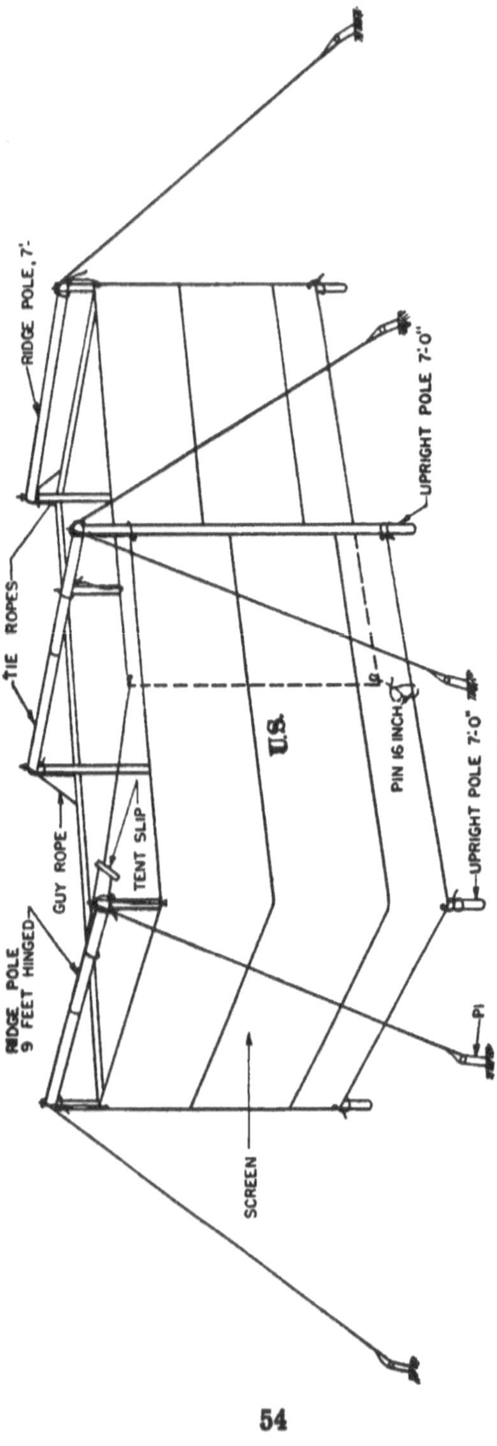

① Side view.

FIGURE 25.—Latrine screen.

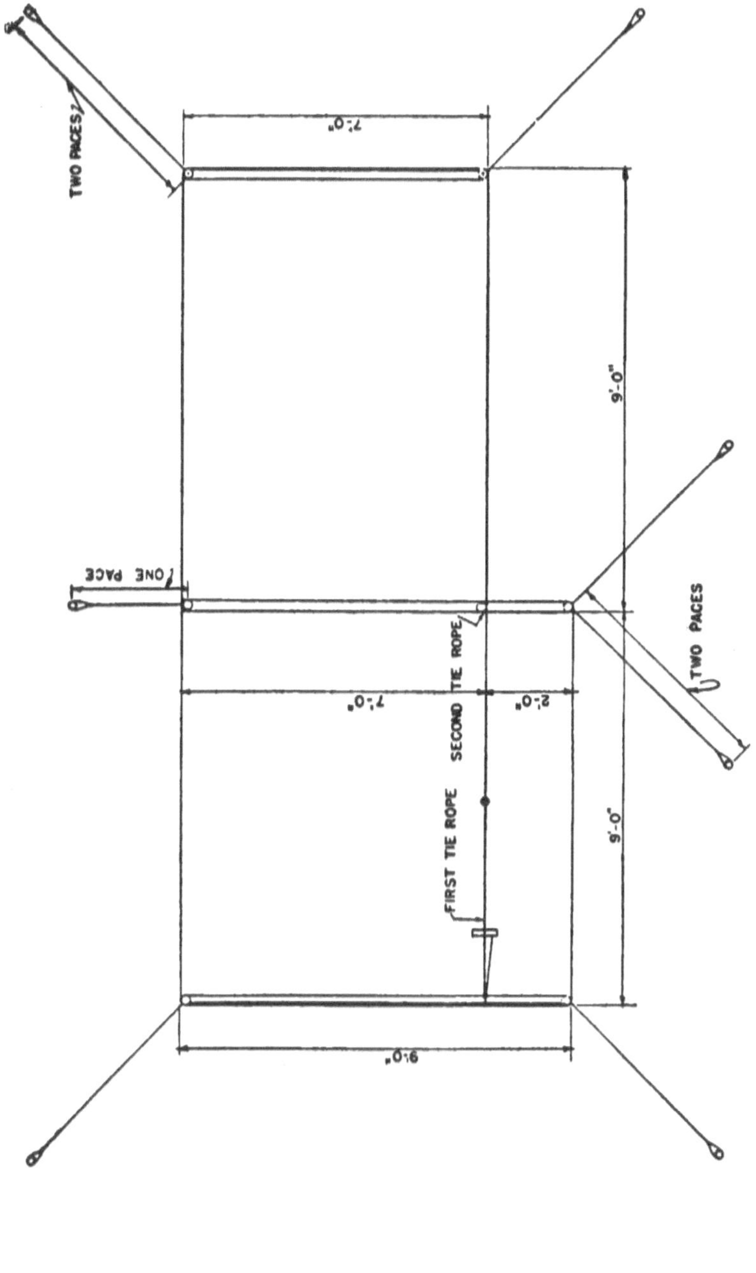

② Top view.

FIGURE 25.—Latrine screen—Continued.

■ 29. STRIKING.—*a*. To strike the latrine screen the work is carried on in the reverse order to pitching the screen. All tie ropes on the lower edge of the screen are untied from the upright poles. The outside of the entrance is untied from the center ridge pole. At the wide end, the latrine screen is untied and dropped to the ground. Guy ropes at this end are detached and the ridge pole and two uprights are disassembled and placed in a pile at one side.

b. The narrow end of the latrine screen is untied. The guy ropes are detached and the ridge pole and uprights are placed with the poles previously placed to one side.

c. The same procedure is followed with the center ridge pole. The eight tent pins and seven guy ropes are collected and placed near the ridge poles and uprights.

■ 30. FOLDING.—*a*. The first step in folding the latrine screen is to extend the screen on the ground and smooth it out. The first fold is made by carrying one end over the other. This step is repeated, the folded edge being placed even with the two ends of the screen. (The screen can be better controlled if the first two folds are made into the wind.) It is important that the canvas be smoothed out after each fold.

b. Next, either the top or bottom edge is folded over one-third of the width of the latrine screen. The other edge is then folded over on top of this fold. The tie ropes and all guy ropes, except one for tying the bundle, are put inside the folds at one end of the screen.

c. To form the final bundle, a two foot fold from each end of the screen is made toward the center. This step is repeated twice leaving the two folded sections two or three inches apart. One folded section is placed on the other and the bundle is ready to be tied.

d. Lay the guy rope over the center of the bundle and turn the bundle over on the rope. The ends of the rope are crossed over the new top of the bundle. The bundle is turned over again on the crossed ropes and tied.

INDEX

INDEX

Made in United States
Troutdale, OR
11/03/2026

56845110R00038